# Studies in History and Philosophy of Science

Volume 70

*Studies in History and Philosophy of Science* is a peer-reviewed book series, dedicated to the history of science and historically informed philosophy of science. The series publishes original scholarship in various related areas, including new directions in epistemology and the history of knowledge within global and colonial contexts. It includes monographs, edited collections, and translations of primary sources in the English language. These cover a broad temporal spectrum, from antiquity to modernity, and all regions of the world.

Elżbieta Jung · Joanna Papiernik ·
Bartosz Żukowski
Editors

# The Forgotten Giants: Overlooked Revolutions in Natural Philosophy

Springer

*Editors*
Elżbieta Jung
Department of History of Philosophy
University of Lodz
Lodz, Poland

Joanna Papiernik
Department of History of Philosophy
University of Lodz
Lodz, Poland

Bartosz Żukowski
Department of History of Philosophy
University of Lodz
Lodz, Poland

ISSN 1871-7381 ISSN 2215-1958 (electronic)
Studies in History and Philosophy of Science
ISBN 978-3-032-34007-8 ISBN 978-3-032-34008-5 (eBook)
https://doi.org/10.1007/978-3-032-34008-5



This Springer imprint is published by the registered company Springer Nature Switzerland AG
The registered company address is: Gewerbestrasse 11, 6330 Cham, Switzerland

# Contents

# Editors and Contributors

## About the Editors

**Elżbieta Jung** is a Professor of History of Philosophy. She holds the position of Chair of the History of Philosophy, and the Director of the Center for the Philosophy of Nature at the University of Lodz. Her research primarily concentrates on the history of natural philosophy and late medieval theology, particularly emphasising the philosophical concepts developed by 14th-century philosophers and theologians associated with the so-called School of Oxford Calculators. Her most recent books are: *Richard Kilvington's "Quaestiones super libros Physicorum." A Critical Edition with an Introduction*, (Lodz, 2025, https://doi.org/10.18778/8331-737-3); with Monika Michałowska, *Richard Kilvington Talks to Thomas Bradwardine about Future Contingents, Free Will, and Predestination. A Critical Edition of Question 4 from "Quaestiones super libros sententiarum"* (Leiden/Boston, 2023); with Robert Podkoński, *Towards the Modern Theory of Motion. Oxford Calculators and the New Interpretation of Aristotle* (Lodz, 2020).

**Joanna Papiernik** is a scholar specialising in Renaissance philosophy and medieval natural philosophy at the University of Łódź, Poland. Her research focuses on the immortality of the soul in Italian Renaissance thought, with particular attention to the works of Marsilio Ficino and fifteenth-century humanist and natural philosophers. She is the author of *Philosophies of the Afterlife in the Early Italian Renaissance: Fifteenth-Century Sources on the Immortality of the Soul* (Bloomsbury, 2024), a study of *Quattrocento* debates on personal immortality, as well as articles on various aspects of Renaissance intellectual traditions. Her research also addresses medieval natural philosophy, especially the Oxford Calculators' interpretations of Aristotelian physics. She is the author (and co-author) of articles on this subject, such as: *How to Measure Different Movements? The Fourteenth-Century Treatise De sex inconvenientibus*, Przegląd Tomistyczny (2019), and *The Influence of Richard Kilvington's Commentary on the Physics on the Anonymous Treatise De sex inconvenientibus*, Recherches de Théologie et Philosophie Médiévales (2025, no. 2).

**Bartosz Żukowski** is an Associate Professor of Philosophy at the University of Lodz. He specialises in early modern philosophy, mainly within the British tradition, with a particular focus on the history and theory of idealism (especially Richard Burthogge, George Berkeley, and Arthur Collier), and ontological pluralism (the Cartesian tradition; John Sergeant). His work also spans theory of knowledge (including non-standard epistemologies, most notably *knowing-by-being*), philosophy of language, and cognitive science. He is the author of *Unnoticed Revolution: Richard Burthogge's Constructive Idealism* (2019, in Polish) and *Esse est percipi? George Berkeley's Metaphysics of Ideas* (2012, in Polish), and co-editor of *Cognition, Meaning and Action. Lodz–Lund Studies in Cognitive Science* (2015). He has translated into Polish George Berkeley's *Philosophical Notebooks* (2007) and Charles H. Kahn's *Language and Ontology* (2008). Most recently, he completed the translation of *The Philosophical Works of Richard Burthogge*, with an extensive scholarly commentary.

## Contributors

**Graziana S. Ciola** Radboud University, Nijmegen, The Netherlands

**Antonio Clericuzio** Roma Tre, Rome, Italy

**Eva Del Soldato** University of Pennsylvania, Philadelphia, USA

**André Goddu** Stonehill College, Easton, MA, USA

**Elżbieta Jung** Department of History of Philosophy, University of Lodz, Lodz, Poland

**Chiara Militello** Università di Catania, Catania, Italy

**Joanna Papiernik** Department of History of Philosophy, University of Lodz, Lodz, Poland

**Moritz René Pretzsch** Department of Philosophy, University of Kassel, Kassel, Germany

**Sylvain Roudaut** CNRS–SPHERE (Paris Cité University), Paris, France

**Joseph Ulatowski** University of Waikato, Hamilton, Aotearoa, New Zealand

**Antonio Vassallo** Faculty of Administration and Social Sciences, Warsaw University of Technology, Warsaw, Poland

**Michał J. Wagner** Institute of Philosophy, Cardinal Stefan Wyszyński University in Warsaw, Warsaw, Poland

**Gustavo Fernandez Walker** FLoV – Göteborgs Universitet, Göteborg, Sweden

**Bartosz Żukowski** Department of History of Philosophy, University of Lodz, Lodz, Poland

# Chapter 1
# Introduction

**Elżbieta Jung, Joanna Papiernik, and Bartosz Żukowski**

This volume emerged as the fruit of the second edition of the international conference "The Forgotten Giants' Shoulders," which took place at the Center for the Philosophy of Nature at the University of Lodz. The conference series, initiated in 2022, represents an ambitious attempt to revise the canon of the history of philosophy of nature—not in order to deprecate the significance of recognised "giants," but to restore the rightful place of thinkers, theories, and concepts that have remained in the shadow of grand historiographical narratives. The main premise of the project is the conviction that the formation of dominant paradigms in Western philosophy of nature was not the work of individual, isolated geniuses, but the result of complicated intellectual processes in which numerous, today often forgotten, scholars were involved.

While the first edition of the conference focused on methodological breaks and continuations in the philosophy of nature, and the third (planned for 2026) promises an exploration of the convoluted ways of mathematization of Western knowledge, the second instalment of the series—of which the present volume is the outcome and which took place in 2024—was devoted to overlooked scientific revolutions in natural philosophy. Although the Kuhnian perspective is tempting, commanding us to perceive the history of natural thought as a series of revolutions led by great figures that emerge like watershed moments from the landscape of the normal course of events, the actual picture proves to be far more complex and nuanced. Not every

E. Jung · J. Papiernik (✉) · B. Żukowski
Department of History of Philosophy, University of Lodz, Lodz, Poland
e-mail: joanna.papiernik@filhist.uni.lodz.pl

E. Jung
e-mail: elzbieta.jung@uni.lodz.pl

B. Żukowski
e-mail: bartosz.zukowski@uni.lodz.pl

 
E. Jung et al. (eds.), *The Forgotten Giants: Overlooked Revolutions in Natural Philosophy*, Studies in History and Philosophy of Science 70,
https://doi.org/10.1007/978-3-032-34008-5_1

breakthrough contribution is immediately noticed as such, nor is a given field of science always favourably disposed toward revolutionary change.

Consequently, some revolutions remained invisible in the shadow cast by a dominating paradigm or influential competing approach, only to be rediscovered after the full exploitation of the theoretical potential of the latter. Other approaches were overlooked due to their philosophical affiliation, historical-cultural context, socio-religious or political background, or because of peculiarities of notation and terminology. Finally, certain innovations were quickly appropriated and developed by more powerful competitors, losing their original identity and authorial attribution.

A clarification of terminology is in order. Throughout this volume the terms "overlooked," "forgotten," and "unnoticed" are employed with distinct, though related, meanings. A theory or thinker may be described as *overlooked* when it was present in the intellectual landscape of its time but failed to attract the attention proportional to its significance—whether because it appeared at an unfavourable moment, in a marginal linguistic or cultural context, or in a form that did not match the dominant institutional idiom. *Forgotten* designates cases of genuine historical erasure: contributions that were once known and discussed, but subsequently dropped out of the record—often as a result of the consolidation of a rival paradigm or the disappearance of the community that sustained them. *Unnoticed*, finally, refers to work that, for structural reasons—disciplinary boundaries, the hermeticism of specialist discourse, or the absence of influential advocates—never achieved the visibility it deserved, even in its own time. These distinctions cut across the volume's chapters in different ways: Autrecourt's *situalitas* is, in the first sense, overlooked; Chałupczyński's selectionism is, in the second sense, forgotten; and certain dimensions of Wittgenstein's engagement with incommensurability have remained, in the third sense, unnoticed within the literature—a point that justifies his inclusion in the present collection alongside figures whose obscurity is of a different, more absolute kind.

The articles collected in this volume take up precisely these issues, exploring both concrete cases of overlooked, forgotten, and unnoticed theories, as well as the broader question of the nature of revolutionary change in natural philosophy and some other branches of philosophy. They represent a broad temporal spectrum: from ancient Greece to the twentieth century, and a diversity of methodological approaches, from detailed philological analysis to conceptual reconstruction and metaphilosophical reflection. They are united by a common conviction that the history of the philosophy of nature is far richer than standard textbook narratives suggest, and that many important concepts and theoretical threads await their discovery or reassessment.

## 1.1 Atomism and its Forgotten Versions

One of the most important threads running through the present volume is the problem of atomism and its various incarnations in the history of the philosophy of nature. Chiara Militello in her article undertakes the challenge of reconstructing the atomistic explanation of a specific physical phenomenon: water evaporation. Although surviving fragments and testimonia of the ancient atomists rarely feature explanations of specific physical phenomena, it is known that these philosophers actually provided such explanations. Militello undertakes a two-stage analysis. In the first stage, she systematically surveys the models of material change that could, in principle, be applied to the phenomenon of water evaporation, identifying six possible explanatory frameworks. In the second stage, drawing on inferences concerning the nature of vapour derived from extant testimonia, she excludes five of these models, thereby isolating an atomistic account as the most plausible explanation of water evaporation. This reflection on a concrete phenomenon provides us with valuable clues as to how the ancient atomists thought about material change and how their theory could function in the practice of explaining natural phenomena.

Gustavo Fernandez Walker analyses the atomism of Nicholas of Autrecourt, focusing on the concept of *situalitas*. Autrecourt, a fourteenth-century philosopher who defended atomism against Aristotelian criticism, introduced this neologism to solve a fundamental problem: how indivisible, dimensionless atoms can compose dimensional objects. Fernandez Walker argues that *situalitas* is not merely an ad hoc solution but constitutes an important element of Autrecourt's atomistic program. This concept refers to the relation that a dimensionless atom maintains with space: it does not occupy a place (*situs*), but possesses its own *situalitas*, which enables it to become part of an extended body. Fernandez Walker shows how this concept allows Autrecourt to bypass the Aristotelian dilemma concerning contact between indivisibles and the problem of making a quantum from non-quanta, thus revealing *situalitas* as far more than a mere conceptual expedient.

Antonio Clericuzio in his article devoted to early seventeenth-century atomists and their conceptions of heat and fire shows how, in the early decades of the seventeenth century, new empirical knowledge together with the revival of ancient atomism prompted philosophers and physicians to reconceptualise theories of heat and fire in corpuscular terms. Two atomists, Isaac Beeckman and Sébastien Basson, abandoned the Peripatetic doctrine of forms and qualities, developing a kinetic theory of heat whereby the shape, size, and motion of material particles explained heat and fire. Clericuzio further demonstrates that, in *The Assayer*, Galileo maintained that heat is nothing more than a designation for the action of minute corpuscles impinging upon the sensory organs. He also notes that Benedetto Castelli, Galileo's student and collaborator, pursued experimental investigations into the relationship between heat and light, research that was subsequently developed by Robert Boyle.

## 1.2 Medieval Logic and Semantics

Graziana Ciola in her extensive study devoted to Marsilius of Inghen, nominalism, and the semantics of necessarily empty terms in the late fourteenth century addresses the key issue of the tension between nominalist commitments and the acceptance of non-existent objects in late-medieval logic. Ciola argues against the common assumption that nominalism's ontological parsimony necessarily excludes reference to *impossibilia* and should even exclude any reference to *possibilia*. Instead, she shows that Marsilius of Inghen and his followers developed sophisticated semantic frameworks that could accommodate empty terms while maintaining their anti-realist commitments in meaningful ways. Through an analysis of supposition theory, ampliation, and the treatment of *imaginabilia*, Ciola demonstrates how Marsilius and his followers created a coherent system that allowed meaningful discourse about non-existent and even impossible objects.

Sylvain Roudaut reconstructs a little-known debate that took place between John Buridan and Albert of Saxony around the middle of the fourteenth century concerning the cause of certain types of motion. The two authors diverged in their accounts of so-called "motions contrary to nature," such as instances in which hot water heats other bodies, despite water being, naturally cold in its essential constitution. Roudaut explains that Buridan regards accidents as the true principal cause of such motions, whereas Albert of Saxony attributes them to the substantial form. The article shows the coherence of Albert's position despite its counterintuitive character and the problems it raises by apparently erasing the distinction between natural motions and violent motions. The reception of this debate by later authors, such as Blasius of Parma and Luis Coronel, illustrates the importance of this question for late medieval physics and shows how some thinkers undermined the fundamental distinctions of Aristotelian philosophy of nature.

## 1.3 Forgotten Revolutions in Cosmology and Astronomy

André Goddu in his article on pre-Copernican astrology at Cracow in 1450–1500 examines the teaching of astrology at Cracow University in the second half of the fifteenth century as background to Copernicus's astrological activity. For over a hundred years scholars rejected any notion that Nicolaus Copernicus practised astrology; it was only the work of Robert Westman from the beginning of the twenty-first century that challenged the orthodox view about Copernicus's exceptionalism in this regard. Goddu summarises the teaching on spherical astronomy and the construction of horoscopes and astrological charts, discusses Copernicus's astrological chart from around 1540, examines Cracow astrologers in the second half of the fifteenth century, discusses the authenticity of Copernicus's annotations on Haly Abenragel's *In judiciis astrorum*, and then turns to the evidence for Copernicus's astrological activity and practice. Contra the exceptionalism attributed to Copernicus's astronomical innovations, Goddu argues that his engagement with astrology was neither exceptional nor anomalous for a Renaissance natural philosopher of his standing, but rather representative of the intellectual conventions and practices endemic to sixteenth-century learned culture, thereby reconceptualizing the relationship between Copernicus's revolutionary heliocentrism and the practical discipline of prognostication: a connection obliquely suggested by the very title of his magnum opus, *De revolutionibus*.

Eva Del Soldato investigates the early modern interpretative reception of Anaxagoras's doctrine of the fiery aether through an analysis of Aristotelian commentaries and the writings of authors who, for philosophical, scientific, and religious reasons, sought to contest Peripatetic cosmology. Her study gives particular attention to Matija Frkić's *De caelesti substantia*, a treatise published in 1646 that endeavoured to establish Anaxagoras as an anti-Aristotelian authority compatible with Christian concerns. Del Soldato assesses the limited influence of Frkić's work and accounts for the reasons why Anaxagoras did not undergo a substantial revival in early modern cosmological discussions, despite certain appealing aspects of his doctrine. She further demonstrates that, although Aristotle's *On the Heavens* briefly attributes to Anaxagoras the view that the celestial aether is in fact fire, early modern interpretations of this passage remained strikingly orthodox, and the critical possibilities inherent in this claim were largely left unexplored.

## 1.4 Polish Contributions to the History of Science

In his article, Michał Wagner examines the figure of Adam Chałupczyński (1822–1887), a Polish physician who, independently of Darwin, formulated an evolutionary theory grounded in the principle of the survival of the best-adapted in the struggle for existence, to some extent anticipating key elements of later Darwinian thought. Chałupczyński developed his theory from 1851 onwards, publishing it in 1860 in *Ideas Toward a Knowledge of the History of Life on Earth* (*Pomysły do wiedzy dziejów życia świata*). Wagner argues that rather than treating his theory as an incomplete or derivative precursor to Darwinism, it should be situated within nineteenth-century natural philosophy and analysed as an original selectionist conception expressed through theological and cosmological language. By reconstructing Chałupczyński's views and their reception, Wagner demonstrates that the idea of natural selection could emerge independently of Darwin and was not confined to British science, supporting the broader claim that selectionism was, in a meaningful sense, "in the air" in the nineteenth century and could assume diverse conceptual forms.

Joseph Ulatowski in his article devoted to Maria Ossowska and the science of morality argues that Ossowska was a forgotten giant of experimental philosophy whose contribution to philosophy paved the way for moral psychology and experimental moral philosophy. Ossowska wrote extensively on moral philosophy, and much of her work focused on the foundations of moral philosophy, including its psychological and sociological aspects. From the turn of the twenty-first century, a branch of experimental philosophy has investigated moral intuitions, judgments, and behaviours using social scientific methods and data to substantiate, challenge, or ameliorate claims of moral philosophy. Ulatowski compares Ossowska's vision of the science of morality with experimental moral philosophy, showing striking similarities between her research program and contemporary empirical work in philosophy. He argues that her empirical approach may have been in part due to her intellectual upbringing in Twardowski's Lvov-Warsaw School, where methodological innovation in philosophical projects was lauded rather than derided.

## 1.5 Contemporary Perspectives on Forgotten Revolutions

Although Wittgenstein himself is among the most extensively studied figures in twentieth-century philosophy, the specific dimension explored in this chapter—the structural parallel between his later philosophy and Kuhn's account of paradigm shifts, and its implications for the natural sciences—has remained at the margins of the secondary literature, making it a genuinely unnoticed contribution within an otherwise well-mapped oeuvre. Pretzsch analyses Wittgenstein's critique of science and technology, maintaining that his fundamental anti-scientistic and anti-technocratic stance retains significant relevance in contemporary contexts. Pretzsch argues that Wittgenstein's scepticism toward causal explanation possesses enduring philosophical importance and that his engagement with what resists explanation is marked by an attitude of awe and respect. The article further contends that a productive and instructive parallel may be drawn between Wittgenstein's later philosophy and Thomas Kuhn's notion of paradigm shifts, particularly insofar as Kuhn's framework resonates with Wittgenstein's concept of a world-picture. In addition, Wittgenstein's account of so-called "deep disagreement" is presented as offering a potentially valuable perspective for the natural sciences. Pretzsch shows that, upon closer examination, Wittgenstein had already developed conceptual resources for addressing problems of incommensurabilities and for facilitating dialogue between different (scientific) world-pictures.

In his article on relational conceptions of space and time, Antonio Vassallo reconsiders the Leibnizian–Machian tradition of relationalism in light of recent advances in physics and metaphysics. Although long regarded as conceptually refined yet technically inadequate, relationalism has recently reasserted itself as a plausible foundation for fundamental physics through the development of pure shape dynamics and the metaphysics of self-subsisting structures. Vassallo maintains that these frameworks respond to enduring philosophical and physical problems concerning empirical adequacy, the reconstruction of physical quantities, and the grounding of dynamical laws. By reassessing the historical development of relationalist thought and elucidating the ways in which contemporary proposals realise its central aims, the article argues for the renewed significance of relationalism within foundational inquiry. It further suggests that what may be described as an "unfinished revolution" remained incomplete because it lacked appropriate conceptual resources: resources that, the author contends, have now been attained.

* * *

Taken together, the studies collected here suggest that the eclipse of a revolution in the philosophy of nature is rarely a matter of chance. Several structural patterns recur across the cases examined in this volume. First, paradigmatic dominance: a powerful and institutionally entrenched framework—Aristotelian physics, Darwinian biology, analytic philosophy of science—tends to absorb or neutralise competing ideas, either by reinterpreting them within its own conceptual grammar or by rendering them simply invisible to the eyes trained on its canonical problems. Second, the mismatch between an innovation and the available reception infrastructure: Chałupczyński wrote in Polish at a moment when the language of evolutionary debate was almost exclusively Anglophone; Autrecourt's *situalitas* appeared in a scholastic idiom already losing its authority; Frkić's Anaxagoras arrived too late for the cosmological debate it sought to enter. Third, what might be called *attribution capture*: ideas that were later developed by more powerful or better-connected thinkers lost their original authorship, becoming absorbed into the intellectual biography of their appropriators. Fourth, the sheer hermeticism of certain specialist domains—late medieval logic, the philology of fragmentary testimonia, shape dynamics in philosophy of physics—which makes genuine innovations effectively invisible to anyone not already embedded in the relevant community. No single pattern explains all the cases discussed here. But the fact that these patterns recur across such chronologically and culturally distant contexts, from ancient Athens to nineteenth-century Galicia to twentieth-century Vienna, suggests that the occlusion of intellectual revolutions is not an anomaly to be explained away, but a structural feature of the history of natural philosophy that deserves sustained critical attention.

The studies collected in this volume, though thematically and methodologically diverse, share a common perspective: the conviction that the history of philosophy is far richer and more complex than dominant historiographical narratives suggest. From ancient atomists through medieval logicians to early modern philosophers and nineteenth-century evolutionists: all these figures and theories remained in the shadow of their better-known successors or contemporaries. Their omission did not result from a lack of originality or intellectual significance, but from contingent historical circumstances, cultural contexts, language barriers, or simply from the fact that their ideas were later formulated more influentially by other thinkers.

# Chapter 2
# Atomistic Explanations of Observable Phenomena: Water Evaporation as a Key to an Overlooked Revolution in Natural Philosophy

**Chiara Militello**

**Abstract** Even though the surviving fragments and testimonia of the ancient atomists rarely feature explanations of specific physical phenomena, it is known that these philosophers actually provided such explanations. Attempts to reconstruct the atomists' description of particular phenomena have been rare, despite the potential for such analysis to provide a more complete understanding of ancient atomism. This chapter aims to partially fill this gap by piecing together the atomistic interpretation of a physical phenomenon, namely water evaporation. In order to deduce how the ancient atomists conceived of the evaporation of water, a two-step analysis is conducted. First, all the models of material change that may theoretically be applied to the case of water evaporation are listed, yielding six different possible explanations of the phenomenon. Then, deductions about the nature of vapour based on the testimonia make it possible to rule out five of the six candidates, thus revealing the likely atomistic explanation of water evaporation.

**Keywords** Democritus · Leucippus · Material change · Elements · Aggregates

C. Militello (✉)
Università di Catania, Catania, Italy
e-mail: chiara.militello@unict.it

 
E. Jung et al. (eds.), *The Forgotten Giants: Overlooked Revolutions in Natural Philosophy*, Studies in History and Philosophy of Science 70,
https://doi.org/10.1007/978-3-032-34008-5_2

## 2.1 Types of Material Change and Water Evaporation

Studying how the ancient atomists[1] explained specific physical phenomena can be useful for a better understanding of their general theories. Specifically, in this chapter I will examine how the atomists interpreted the phenomenon of water evaporation. It is likely that they explained this phenomenon in terms of their atomistic theory. We know that Democritus provided an atomistic interpretation for every meteorological phenomenon he discussed (see Wilson 2013, 22–24), and the atomists discussed the evaporation of water. We know this from a text where Aristotle says that, according to Democritus, the volume of the sea was progressively diminishing, a process that would not stop until the sea had progressively disappeared.[2] Although this brief testimony does not explicitly state how the drying up of the sea occurs, it is possible to deduce it from Aristotle's response, which immediately follows the testimony. The objection Aristotle makes to Democritus and all other thinkers who thought the sea was progressively drying up begins with the premise that they must accept one of two contrary propositions: either the water drawn up (*anagesthai*) by the Sun does not return in the form of rain, or it does (Aristotle, *Meteorologica* 2.3356b21–24).[3] Evidently, Democritus believed that the reason the volume of water was decreasing was evaporation; otherwise, Aristotle could not have responded that he needed to explain what happened to the evaporated water.

Given that the atomists provided an atom-based[4] explanation for all the meteorological phenomena they discussed, and that water evaporation was one of these phenomena, it is likely that they had described how water evaporates in terms of their theory. Since no fragment or testimony about this explanation has survived, we can try to reconstruct what this explanation was, based on the fragments and testimonies that have survived. In these texts, there are four types of material change that could apply to the case of water evaporation. I will dedicate a subsection to each type of change below. For each type of change, the atomists provide one or more explanations in terms of atoms; therefore, at first glance, we can consider all these

---

[1] I speak of "the ancient atomists," meaning Leucippus and Democritus without distinction, since it is not possible to clearly distinguish their theories (see Laks and Most 2016, 7:3). That said, since I am discussing the atomistic explanation of a specific phenomenon like evaporation, it is likely that this explanation was developed by Democritus and not Leucippus, given that the latter probably provided only the general system of atomism, which Democritus then specified (see Laks and Most 2016, 7:3). I consider primarily the fragments and testimonies of the atomists collected in Laks and Most (2016, vol. 7), which is also the edition I cite from (abbreviated as: LM 27, when citing a fragment or testimony; in all other cases, i.e., when referring to the editors' opinion, I cite the work as Laks and Most 2016, 7. The further indispensable literature on the atomists is Luria (1970), Morel (1996), Taylor (1999), Brancacci and Morel (2007), Leszl (2009).

[2] LM 27 D122.

[3] Laks and Most (2016, 7:406–408) consider only 356b9–15 as Aristotle's response to Democritus, but it is clear that in the subsequent lines, Aristotle also has Democritus among his polemical targets.

[4] For terminological consistency, in this essay I will always speak of "atoms," even though other terms are also found in the fragments and testimonies ("first bodies," "principles," "shapes," "elements," "substances," "seeds": see Leszl 2009, Quadro sinottico, 19).

explanations as candidates for the explanation of water evaporation. I will indicate each of these explanations with a letter, from A to F.

### *2.1.1 The Generation of One Element from another*

Water evaporation could be the apparent generation of one element[5] (air) from another (water). One reason to think that evaporation was conceived in these terms by the atomists is that Aristotle mentions evaporation in his response to the conception of Empedocles and Democritus on the mutual generation of elements (Aristotle, *De Caelo* 3.7305a33–b26).[6] Furthermore, the identification of vapour with air was a widely held view at the time of the atomists (see Leszl 2009, 34).

For atomists, the generation of one element from another is *apparent* because the atoms that form elements of different types have different characteristics. Apart from fire, whose atoms have a spherical shape (LM 27 D59, D60b, D132), the difference between the elements, and in particular between water and air, is given by the magnitude of the atoms (LM 27 D59, R33). We also know that atoms are immutable (LM 27 D43, D44, R88). Since different elements are composed of atoms of different magnitudes, and the magnitude of atoms does not change, one element cannot be generated from another. When it seems that element Y is generated from another element X, what happens instead is that atoms of type Y, which were previously mixed with those of type X, separate from them, as is explained in testimonium R33:

> For it is not possible that they [air, earth, and water] come to be out of each other; for the largest bodies, in separating (*enkrinesthai*), will always be lacking (*hupoleipein*)—but it is in this way that they [Leucippus and Democritus] say that water, air, and earth come to be out of each other. (LM 27, 393)[7]

Here we have the first candidate for the explanation of water evaporation:

(A) if the evaporation of water is conceived as generation of air from water, it is nothing but the separation of air atoms from the water atoms with which they were mixed.

One might think that this explanation does not apply to evaporation, because Aristotle is talking about the separation of larger atoms from smaller ones, and air atoms are probably smaller than water atoms. However, this objection is not valid, for two reasons. First, Aristotle says that air is also generated, and in general that the elements

[5] In this essay, I use "elements" to mean earth, water, air, and fire, even though in some testimonies the atoms themselves are called "elements" (see Leszl 2009, Quadro sinottico 19).

[6] This text is not included in LM 27, but is testimonium 55a in Taylor 1999.

[7] Aristotle's criticism of this doctrine is that at a certain point, there will be no atoms of the generated element left in the generating element. This is problematic because it does not correspond to what we see, as the generating element cannot completely transform into the generated element (see Aristotle 1922, ad locum; Aristotle 1939, 293) and/or at a certain point an element will stop producing other elements (see Kouremenos 2013, 90; Aristotle 2020, 223).

are generated from one another. Second, in a passage reminiscent of R33, Aristotle speaks of elements with smaller atoms being generated from those with larger atoms (Aristotle, *De Caelo* 3.7305a33–b26).[8] Therefore, the doctrine of R33 can be applied to the generation of air from water; indeed, Stocks (Aristotle 1922, ad locum) and Guthrie (Aristotle 1939, 293) mention exactly this phenomenon in their explanation of the passage.

On this view, evaporation could be the separation of air atoms from the water atoms with which they were mixed. It can be noted that, if for the atomists atoms had weight, this explanation overlaps with that of the upward expulsion of a lighter body from a heavier one (LM 27 D49). However, it is not clear from the sources whether atoms had weight (LM 27 D48–D51). Furthermore, one can hypothesise that the cause of the separation of air atoms from water atoms is that atoms tend to move towards similar atoms (LM 27 D32, D55, D72, D123). Vapour would consist of air atoms separating from the water atoms with which they were mixed to join other air atoms.

Two things remain to be clarified regarding the generated element: first, is it pure (i.e., is vapour pure air)? And second, is it generated in the form of single atoms or an aggregate[9] (i.e., is vapour the name we give to separated atoms, or is it an aggregate)?

I do not believe there is a way to answer the first question. Indeed, Aristotle's description of the generation of element Y from element X is compatible with two options: 1) from a mixture X + Y (with a majority of X, as it appears to be X), Y separates; 2) from a mixture X + Y with a majority of X, a mixture X + Y with a majority of Y separates (as it appears to be Y).

It should be noted here that the fact that vapour condenses into water does not exclude the possibility that vapour is only air. One might think that the fact that vapour condenses into water excludes the possibility that vapour is only air, because to condense into water, it must contain water atoms. However, this model provides that atoms of different magnitudes can not only separate, but also unite, as seen in various fragments stating that atoms "gather together" and "become interlaced" (LM 27 D29, D30, D80, D81). This would allow an explanation of how rain, which is water, can be "generated" from vapour, if vapour is air. After evaporation produces a vapour composed solely of air atoms, water atoms could join this vapour; these water atoms could later separate from the air atoms, producing rain.

It is also not clear whether, in the apparent generation of an element, single atoms or an aggregate separate, or if, in different cases, both possibilities occur. Aetius attributes to Leucippus the thesis that thunder is the expulsion of fire from within a cloud, and to Democritus the thesis that thunder is an aggregate found inside a cloud (LM 27 D117). This might suggest that for the atomists, thunder occurs when an aggregate of fire separates from another element (the one of which the cloud is composed). However, in Leucippus' case, there is no mention of an aggregate (nor is it clear if the phenomenon falls under the type of apparent generation of one

[8] This text is not included in LM 27, but is testimonium 55a in Taylor 1999.

[9] In this essay, for linguistic consistency, I always refer to visible bodies, composed of atoms, with the term "aggregates."

element from another), and in Democritus' case, the aggregate is non-homogeneous (*anomalos*), so it does not only include fire. Therefore, this passage does not provide us with any information on the status of the atoms of the element that seems to be generated from another element.

### 2.1.2 The Change of State of an Aggregate

Water evaporation can also be conceived as the transition of an aggregate (water) from one "state" (to use a modern expression) to another (from liquid to "gaseous," again using an anachronistic expression). Such a process is explained in testimonium R47[10]:

> In general, it is absurd that this [i.e., acting and being acted upon] happen only in this way, namely when bodies are split apart. For this thesis abolishes alteration (*alloiôsis*), but we see that the same body, even though remaining continuous, is at one time liquid, at another solid, and that it undergoes this affection not because of separation and reunification, nor because of "turning" (*tropê*) and "disposition" (*diathigê*), as Democritus says. For it is not because there has been a change of disposition (*metatassesthai*) or of placement (*metatithesthai*) in its nature that it has changed from being liquid to becoming solidified, nor do there now [i.e., in this state] exist within it hard and solidified bodies indivisible in their masses; but just as it is entirely liquid, it is sometimes hard and solidified." (LM 27, 405–407, with one modification)

Aristotle is asserting that Democritus' theories fail to explain the phenomenon of alteration, i.e., the change in a subject's qualities (but not its substance). As an example of alteration, Aristotle uses the transition of an aggregate from solid to liquid or vice versa. As Crubellier (2004, 287–288) explains, Aristotle is saying that for Democritus this transition can occur in two different ways. On the one hand, when the solid atoms separate from the aggregate, it is no longer solid; when they rejoin it, it is solid again. On the other hand, when the atoms in the aggregate change their order or position,[11] it can become liquid; when they change them again, it can become solid again. Aristotle replies that the change of state cannot occur through separation or reunification, because it is always the same body, without loss or addition, that is at one moment liquid and at another solid. Nor can it occur because the individual atoms change order or position, given that (1) it is a change that affects the object in its entirety, not its individual parts, and (2) in the case of state changes like the freezing of a liquid, the change is instantaneous, which cannot be true of the change in order or position of a multiplicity of atoms. The fact that Democritus cannot explain the

[10] Turning and disposition (or the Aristotelian equivalents, i.e., position/*thesis* and order/*taxis*, see D31 and D32) are cited as possible causes of alteration/*alloiôsis* also in D56 and D72.

[11] It may be useful to recall the concepts of *tropê* and *diathigê*, with the help of LM 27 D31. *Diathêkê* or *diathêgê* or *diathigê* (see Laks and Most 2016, 1:228), i.e., disposition (Laks and Most's translation)—*taxis*/order in Aristotle's terms—is the difference between the combinations of letters "AN" and "NA." *Tropê*, i.e., turning (Laks and Most's translation)—*thesis*/position in Aristotle's terms—is the difference between the letters Z and N.

transition of a body from liquid to solid and vice versa shows, according to Aristotle, that Democritus' theories fail to explain alteration.

Note that, even though one of the arguments against one of the explanations provided by Democritus concerns an instantaneous change (freezing of a liquid), Aristotle attributes these two theories to Democritus as explanations for non-instantaneous changes as well (transition from solid to liquid state). Therefore, nothing prevents us from thinking that evaporation (which is not instantaneous) could have been explained by Democritus in one of the two ways mentioned. The first explanation of state change could, in turn, be applied to the evaporation of water in two different ways.

(B) One can conceive of the evaporation of water as a separation of liquid atoms from gaseous ones (condensation would be the reunification of these atoms). When the liquid atoms separate, the aggregate is no longer liquid.

In a sense, this explanation of water evaporation is fundamentally the same as explanation A, because in both cases, it involves conceiving of water evaporation as the separation of atoms of a certain type (air, gaseous) from atoms of another type (water, liquid). Since individual atoms are characterised only by shape and magnitude,[12] "atoms of a certain type" means "atoms of a certain shape and/or magnitude." Thus, explanations A and B can be combined by saying that atoms of a certain shape and/or magnitude separate from atoms of a different shape and/or magnitude. In the case of explanation A, we know that air atoms are distinguished from water atoms by magnitude.

A difference between the two explanations is that in explanation A, it is the air atoms that separate from the water atoms, whereas in explanation B, the liquid atoms (corresponding to the water atoms of explanation A) separate from the gaseous ones (corresponding to the air atoms). This difference could be crucial, given that it is unclear in explanation A whether the vapour that separates from the water is an aggregate or the name we give to a multitude of single atoms, whereas in explanation B, the vapour is definitely an aggregate—it is the aggregate from which the liquid atoms separate.

(C) However, one can also conceive of the evaporation of water as the union of gaseous atoms with liquid ones (in this case, condensation would be the separation of gaseous atoms from liquid ones). When gaseous atoms unite with liquid ones, the aggregate becomes gaseous.

(D) Applying the second mode of state change to the evaporation of water, one would conceive of evaporation as a change in the order and/or position of the water atoms.

[12] The other distinctive characteristics are order and position, which are relative not to the single atom but to its relationships with other atoms. Texts about the differences between atoms: LM 27 D45 (shape); D29, R15, R33, R64, R94 (shape and magnitude, also mentioned among other features in R34); D31, D32 (shape, order and position).

Explanation D is the view adopted by the only scholars who have referred to the atomistic conception of evaporation: van Melsen and Taylor. Notably, they make no reference to fragments and testimonies.

It may be worth quoting their words, starting with van Melsen (1952, 21):

> To begin with the states of aggregation, evaporation consists in the loosening of atomic connections. Therefore, vapour will occupy more space than the liquid before its evaporation. The difference between a liquid and a solid is likewise reduced to a shift in position of the atoms.

Van Melsen explicitly assimilates the case of evaporation to that of the transition from liquid to solid or vice versa, which, according to him, is always a change in the position of atoms. Specifically, when a liquid evaporates, its atoms, while remaining bound to each other, become more distant from one another.

Taylor (1999, 85, n. 78) expresses a similar view:

> Aristotle here returns to the objection to atomism raised in 54a, viz. that on the atomists' theory the elements, earth, air, fire, and water, are not genuinely transformed into one another, as in Aristotle's own theory, but separated out from a pre-existing mixture. The atomists reject Aristotle's premiss that the elements are stuffs of irreducibly different kinds: on their view the elements, i.e., the basic substances, are the atoms, which are uniform in kind, and earth, air, etc. are differently structured aggregates of atoms. Their account of what it is, e.g., for water to turn into air (in evaporation) is that one comparatively stable structure is transformed into another, less stable one, perhaps by the breaking up of some of the bonds linking the atoms in the former (no details of any atomistic accounts of such processes survive).

Here, Taylor is commenting on Aristotle, *De Caelo* 305a33–b26, a text that is not included by Laks and Most among the testimonies on the atomists. In this passage, Aristotle discusses the atomistic doctrine of transition from one element to another, i.e., the doctrine I focused on in Sect. 1.1. In this context, Aristotle cites evaporation as the apparent generation of air from water.[13] According to Taylor, the atomists interpreted evaporation and similar phenomena as the change of structure of an aggregate. Water and air are aggregates of atoms of the same type, but structured differently; when the structure of an aggregate changes from water-like structure to air-like structure, evaporation happens.

That earth, air, and water are aggregates of the same atoms, just structured differently, seems to be contradicted by Aristotle, *De Caelo* 303a3–b8, a passage that includes testimonia D59 and R33. In this passage, as I mentioned earlier, Aristotle says that the different elements are distinguished based on the magnitude of their atoms. *De Caelo* 303a3–b8 is testimony 54a in Taylor's numbering, and in the passage I quoted, Taylor cites this testimony in relation to the thesis that the generation of the various elements from one another is only apparent. However, he does not cite it in relation to the differences between the atoms of the various elements—either here, in the notes to the testimony (Taylor 1999, 82–83), or when he refers to it in his commentary (1999, 168, 169.). Taylor (1999, 173–175, 182, 211–213) admits that atoms of various magnitudes (he uses the word "size") exist. He also admits that magnitude is one of the basic properties of atoms (1999, 75, n. 63; 175, 179,

---

[13] Earlier, I cited this passage as a possible proof that the atomists considered vapour to be air.

180, 194), and that for this reason different atoms act differently (1999, 167, 172, 177, 183, 193; cf. 179–180). Evidently, he does not believe that the effect of these differences is the creation of an aggregate of earth rather than water or air. The reason is probably that he thinks that, for the atoms forming aggregates to connect with each other, it is necessary for them to be of different magnitudes[14] (1999, 173).

Regardless, based on the idea that all elements are made of atoms of the same type, Taylor reduces the separation of one element from another to a change in the relationships between some atoms. He states that no details of how this process occurs have survived. If he means that there are no testimonies on "changes of state," I disagree, because there is the passage I quoted at the beginning of this subsection, i.e., R47. Taylor cites this passage[15] as a "further reference" with respect to testimony 45 (LM 27 D32). In the latter passage, from Simplicius' commentary on *Physics*, there is no clear reference to states of matter, but when Theophrastus (whom Simplicius is quoting) says that each of the forms, arranged in a different aggregate, produces a different *diathesis*, Taylor (1999, 72) translates the latter word as "state," and he may mean states of matter such as solid and liquid. But perhaps Taylor, when he says that no details remain on how the change in relations between atoms occurs, means that there are no testimonies on what "turning" and "disposition" precisely mean and on what a change in the position or order of atoms really is. This would explain why Taylor presents the idea that the apparent generation of elements is due to a breaking of some bonds between atoms as a mere hypothesis.

### *2.1.3 The Diminution of an Aggregate*

Water evaporation can also be conceived as the diminution of an aggregate (water). Generally, the atomists thought that the diminution of an aggregate occurs through the emission of atoms. The relevant testimonium is D30,[16] where Aristotle states that, according to Leucippus, "just as Empedocles and some of the others say that there is affection by means of passages, in the same way all change and all affection are produced in this way, dissociation and destruction coming about by means of the void, and similarly growth, when other [*scil.* atoms] penetrate" (LM 27, 91, with one modification). If the growth of an aggregate occurs through the addition of new atoms, its diminution occurs through the emission of some of its atoms.[17]

[14] And shapes.

[15] More specifically, he cites Aristotle *De Generatione et Corruptione*, I.9327a.16–21 (R47 extends from line 14 to 26).

[16] This passage is part of testimony 48a in Taylor's numbering. Taylor refers to the thesis that aggregates grow by addition of atoms (nor, for that matter, to the implication that the diminution of an aggregate is due to some atoms detaching from it) neither in the notes to the passage (1999, 73–76) nor when he cites the testimony within his commentary (1999, 157, 160–161, 163, 166, 172–173, 174, 179, 184, 219–220).

[17] Note the difference between this model and the model of the corruption of an aggregate by the dispersion of its atoms discussed in D29. While in D29 the aggregate is *totally* destroyed, *all* its

(E) Applied to the evaporation of water, this model would lead one to say that evaporation is the emission of atoms from a pool of water.

The difference with respect to explanation B is that in explanation B, the vapour is formed by an aggregate of gaseous atoms, whereas in explanation E, it consists of single water atoms.

### 2.1.4 *Effluxes*

The evaporation of water could be an efflux that separates from a pool of water. Theophrastus says that according to Democritus, some efflux (*aporroê*) is given off from all things (LM 27 D147). From this passage, one might deduce that these effluxes are simply the matter that allows for perception at a distance, and indeed, effluxes are cited to explain perception at a distance in other testimonies.[18] However, in the *Quaestiones* attributed to Alexander of Aphrodisias, it is stated that Democritus uses effluxes (*aporroiai* and bodies that move *aporreonta*) to explain why a piece of iron moves towards a magnet (LM 27 D126). Therefore, for Democritus, effluxes are generated from every body, and they do not have the sole effect of making vision possible. On this basis, one might think that the evaporation of water could also, in principle, be an efflux.

However, the most parsimonious generalisation of the testimonies on effluxes is the one proposed by Leszl (2009, Quadro sinottico, 26): effluxes explain the interactions between two aggregates (the seen object and the seeing subject, or iron and magnet). Therefore, if evaporation must be thought of as an efflux, one must understand what aggregates interact through it. Considering that Democritus says that the Sun feeds on vapour (LM 27 D224), one can reach the following conclusion.

(F) The evaporation of water could be an efflux that explains the interaction (nourishment) between the "pool of water" aggregate and the "Sun" aggregate.

An efflux is composed of the atoms of the aggregate from which it derives; therefore, in this explanation, vapour is made of water atoms. We must, therefore, compare this explanation with the other two in which vapour is formed by water atoms: D and E. The fundamental questions to ask to make this comparison are: (1) is the efflux a modification of the position of atoms, a separation, or something else? (2) is the efflux composed of separate atoms or is it an aggregate? The answer to the first question may not be attainable, as it does not seem possible to deduce the exact mechanism of effluxes from the surviving fragments. As for the second question, considering that Democritus says that an efflux can contain something[19] (LM 27 D66), it seems that

atoms separating from one another, in D30, besides this phenomenon, there is also (implicitly) that of *partial* diminution, due to the separation of *some* atoms.

[18] The words used to indicate the efflux are *aporroê* (LM 27 D66, R57, R59 three times), *aporroia* (27 D66, D146) and *aporreon* (27 D67, R25).

[19] Specifically, air.

effluxes are aggregates. On this basis, explanation F could coincide with D (because in both cases vapour is an aggregate of water atoms), but this is not necessarily so (because, while in explanation D evaporation is definitely a modification of the position of atoms, in F the mechanics of evaporation are not clear).

## 2.2 The Atomistic Explanation of Water Evaporation

Now we have a picture of the six explanations that the atomists *might* have conceived for water evaporation. I have recapitulated these explanations in Table 2.1.

We can now try to determine which explanation the atomists *actually* subscribed to. To do so, we can proceed by elimination, removing the explanations that are incompatible with the statements of the atomists—specifically, with their statements about the nature of vapour. The remaining explanation(s) will have a good chance of being the one(s) actually adopted by the atomists. To this end, we must investigate how the atomists answered the following two questions: (1) what type of atoms is vapour composed of? (2) is it composed of single atoms or is it an aggregate?

### *2.2.1 The Nature of the Atoms Composing Vapour*

The opinion of the atomists that allows us to determine what was, according to them, the nature of the atoms composing vapour is that total evaporation of a pool of water is possible. One might think that total evaporation is an obvious fact. Indeed, Baldes (1972, 184–185) believes that for Aristotle this is the case: "for instance, a given volume of water cannot change completely into air (vapour)—a change which

**Table 2.1** The atomists' explanations of water evaporation

| Name | Type of process | Type of atoms forming the vapour | Nature of the vapour |
|---|---|---|---|
| A | Separation of atoms of one type from atoms of another type | Air, or air and water with a majority of air | Aggregate or single atoms? |
| B | Separation of atoms of one type from atoms of another type | Gaseous atoms | Aggregate |
| C | Union of atoms of different types | Water and air | Aggregate |
| D | Modification of the position of atoms | Water | Aggregate |
| E | Separation of single atoms | Water | Single atoms |
| F | Efflux | Water | Aggregate |

quite obviously takes place."[20] I do not think that we can automatically consider total evaporation an obvious fact for the atomists, as they could have thought that, even when such a phenomenon apparently takes place, an invisible quantity of water remains. However, we know that, in at least one case, the atomists believed that evaporation actually leads to the total disappearance of water. As I have explained in Sect. 1, the atomists believe that the sea will evaporate completely.

If vapour were made of air or gaseous atoms, the total evaporation of the sea would imply that the sea never contained water or liquid atoms, which is impossible. This rules out two explanations: the version of A that has vapour composed exclusively of atoms of air, and B. However, the other version of A (the one according to which vapour is composed of air and water, with the atoms of air prevailing in number) is also excluded, because if this explanation were true, the eventual total evaporation of the sea would imply that the sea is composed of a majority of air atoms, which seems implausible. The total evaporation of the sea rules out explanations A and B.

### *2.2.2 Vapour as Single Atoms or as Aggregate*

In the two testimonies where the formation of clouds is discussed, there are clues that lead one to think that vapour consists of atoms not bound to each other, although we cannot be certain.

In a testimony from Aetius, it is said that for Democritus, when vapour forms clouds, it does so by the effect of *pilesthai* (LM 27 D120A). *Pilein* is a synonym for "to assemble"/*sunkrinein* and the opposite of "to dissociate"/*diakrinein* and "to divide"/*diairein* (LM 27 D57a). To assemble (again, *sunkrinein*) means to generate an aggregate from single atoms (LM 27 D13, D29). Therefore, before being assembled into a cloud, the atoms that formed the vapour were separate from one another.

In the other testimony where cloud formation is discussed, Diodorus Siculus says that according to Democritus, clouds are produced (*gennasthai*) in regions where there is a large exhalation of vapours[21] (LM 27 D120b). In all other testimonies where *gennan* is used in relation to inanimate objects,[22] this verb indicates the production of an aggregate from single atoms.[23]

[20] This is, in general, Aristotle's reply to the atomistic theory of the mutual generation of elements: it cannot explain cases where it is evident that one element completely transforms into another.

[21] The term used here is *anathumiasis*.

[22] In testimonies where *gennasthai* is used in relation to living beings, including another testimony preserved by Diodorus Siculus (LM 27 D129), this verb is used to indicate the generation of the living being or one of its parts (27 D129, D172, D175, D202). *Gennan* is then used metaphorically in the maxim, attributed to Democritus but whose authorship remains doubtful, in which it is said that forgetting one's own errors generates impudence (27 D272).

[23] Democritus believes that atoms produce (*gennan*) aggregates (*ta sunkrimata*), i.e., earth, water, fire, and air, which are assemblages (*sustêmata*) of atoms (LM 27 D13). According to Democritus, atoms produce (*gennan*) and form by aggregation (*sunkrinein*) visible objects (but do not produce a single nature, i.e., they do not merge) (27 D29). According to Leucippus, atoms produce (*gennan*)

In both testimonies where the formation of clouds is discussed, it is said that they are formed from vapour, and this process is indicated with verbs that, in the testimonies on the atomists, are normally used for the formation of an aggregate from single atoms. This suggests that clouds, too, are formed from single vapour atoms. If two different authors, in paraphrasing the atomists' thought on cloud formation, use verbs that, when speaking of the atomists, indicate the formation of objects from single atoms, it could be a sign that the clouds are formed from single atoms.[24] Of course, since these are testimonies and not fragments, no definitive conclusions can be reached on this matter. If we had fragments on both the formation of clouds and the production of aggregates (and if, of course, the same verb were used in these fragments), we could be more certain in asserting the coincidence of these two processes.

We can therefore hypothetically conclude that vapour is constituted, for the atomists, of single atoms and is not an aggregate. This rules out explanations B,[25] C, D, and F, which all implied that vapour was an aggregate.

### *2.2.3 Conclusion: Evaporation as Emission of Single Atoms of Water*

The analysis of the nature of the atoms composing vapour and of their possible links has ruled out all explanations but E. Therefore, we can conclude that, for the atomists,

---

by combining (the verb used is *suntithenai*) and intertwining (*periplekesthai*) with each other (27 D30). Leucippus and Democritus produce (*gennôsi*) all other things from the differences between atoms (27 D31). According to Leucippus and Democritus, all things are generated (*gennasthai*) through the intertwining (*sumplokêi*) and combination (*epallaxei*) of atoms (27 D45). It should also be noted that the fire atoms which according to Democritus are the cause of lightning and thunder are called *ta gennêtika tou puros* (27 D117). It must be said, however, that in the same testimony where it is said that clouds are produced where there is much evaporation, it is also said that clouds, when they are destroyed (*thrauomena*), produce (*gennan*) rains (27 D120b). This occurrence of *gennan* can also be interpreted as a production from single atoms, if one conceives the destruction of the cloud as a reduction to single atoms, which then unite into raindrops; but this is only a hypothesis.

[24] Since these are testimonies and not fragments, we cannot say that the atomists used the verbs *pilesthai* or *gennasthai* to indicate the process of cloud or aggregate formation. However, it could be the case that these were technical terms of the atomists, given that they are used in various testimonies. This kind of hypothesis is not uncommon in literature. For example, Taylor (1999, 83, n. 75) states that *epallaxis* could be an atomistic term based on the fact that it is used in a testimony. He also attributes to the atomists the expression "what it is" to indicate atoms (1999, 217), and (at least provisionally) the three synonyms 'nothing,' 'what is not', and 'the void' to indicate what separates atoms (1999, 184) on the basis of testimonies alone. Also, both Taylor (1999, 211–216) and Vassallo (2018, 117) attribute to Democritus the idea that gods are *eidôla* based only on testimonies. Finally, McDiarmid (1958, 294) considers the frequent use of compounds of *plekein* within the doxographies on the atomists as an indication that those terms were used by the atomists. However, here I do not want to argue that *pilesthai* and *gennasthai* were used by the atomists.

[25] Of course, explanation B is also ruled out by the consideration about total evaporation presented in the previous subsection.

the evaporation of water was probably the diminution of water by the emission of its atoms.[26]

This conclusion contradicts the previous literature on evaporation in atomism, i.e., van Melsen and Taylor, who thought that for the atomists, evaporation was a change in the structure of an aggregate.[27] I think that the hypothesis of these two scholars is refuted by the fact that vapour consists of single atoms, and not an aggregate of atoms bound together (albeit by looser bonds than when they were actual water). That vapour consists of single atoms can be deduced from the fact that the testimonies say that clouds are formed by aggregation from vapour, which implies precisely that vapour consists of single atoms.

In this chapter, through an exhaustive analysis of the different models of material transformation, I have reached an original result regarding the ancient atomists' conception of water evaporation. This study also suggests a new avenue for research, because if one were to study how the atomists explained other specific phenomena like the evaporation of water, one could reconstruct the atomists' philosophy of nature in detail.

## References


Aristotle. 1922. *On the Heavens*. Translated by J. L. Stocks. Oxford: Clarendon Press.

Aristotle. 1939. *On the Heavens*. Translated by W. K. C Guthrie. Cambridge, MA: Harvard University Press.

Aristotle. 2020. *De Caelo*. Translated by C. D. C. Reeve. Indianapolis: Hackett.

Baldes, Richard W. 1972. *Aristotle's Relation to Democritus Reconsidered and Vindicated as Against the Criticism of Harold Cherniss. PhD diss.* Chicago: Loyola University Chicago.

Brancacci, Aldo, and Pierre-Marie Morel, eds. 2007. *Democritus: Science, The Arts, and the Care of the Soul*. Leiden: Brill.

Crubellier, Michel. 2004. On Generation and Corruption I. 9. In *Aristotle's On Generation and Corruption I*, edited by Frans de Haas and Jaap Mansfeld, 267–288. Oxford: Clarendon Press.

Kouremenos, Theokritos. 2013. *Aristotle's de Caelo Γ*. Stuttgart: Franz Steiner.

Laks, André, and Glenn W. Most. 2016. Early Greek Philosophy. In *Loeb Classical Library*, vol. 9 vols. Cambridge, MA: Harvard University Press.

Leszl, Walter. 2009. *I primi atomisti*. Firenze: Leo S. Olschki.

Luria, Salomo. 1970. *Democritea*. Leningrad: Nauka.

McDiarmid, J. B. 1958. Phantoms in Democritean Terminology: *ΠΕΡΙΠΑΛΑΞΙΣ* and *ΠΕΡΙΠΑΛΑΣΣΕΣΘΑΙ*. *Hermes* 86 (3): 291–298.

Morel, Pierre-Marie. 1996. *Démocrite et la recherche des causes*. Paris: Klincksieck.

Taylor, C. C. W. 1999. *The Atomists: Leucippus and Democritus. Fragments*. Toronto: University of Toronto Press.

van Melsen, Andrew G. 1952. *From Atomos to Atom*. Pittsburgh: Duquesne University Press.

Vassallo, Christian. 2018. Atomism and the Worship of Gods. *Philosophie antique* 18:105–125.


[26] This is essentially the current notion of evaporation. Therefore, Democritus anticipates not only the idea of atoms, but also some aspects of the behavior of atoms and, in general, of matter.

[27] See the analysis of hypothesis D in Sect. 1.2.

**Chiara Militello** is Associate Professor of Ancient Philosophy at the University of Catania. Her research explores the history of ancient philosophy ranging from the Presocratics to the late Neoplatonists, with a specialised focus on the tradition of the commentators on Aristotle. She has served as a Visiting Scholar at New York University and currently serves on the Board of Directors of the International Society for Neoplatonic Studies (ISNS). At the University of Catania, she acts as Principal Investigator for the departmental team on the project "From Object to Text." Prof. Militello's translation of *Olympiodorus: On Aristotle Meteorology 2–3* has been published in the *Ancient Commentators on Aristotle* series (2026). She is the author of four monographs, including studies on the commentaries on Porphyry's *Isagoge* (2010), the commentary on Aristotle's *De anima* traditionally attributed to Simplicius (2013), Aristotle's dialectic (2017), and Alexander of Aphrodisias's commentary on Aristotle's *Topics* 4 (2017).

# Chapter 3
# A World without Violence? Heterodox Theories of Natural Motion in the Fourteenth Century

**Sylvain Roudaut**

**Abstract** This article offers a reconstruction of a debate that took place between John Buridan and Albert of Saxony around the middle of the fourteenth century concerning the cause of certain types of motion. More precisely, these two authors disagree on how to explain "contrary-to-nature" motions, such as the case of hot water heating, even though the nature of water inclines it instead to be cold. The article aims to explain the disagreement between Buridan and Albert of Saxony and to assess the consequences of their respective solutions. After introducing the problem (Section 1), the theoretical background necessary for understanding the debate is explained (Section 2). It is then shown that Buridan regards accidents as the true principal cause of such motions, whereas Albert of Saxony attributes them to the substantial form (Section 3). A comparison of the two solutions is subsequently offered, showing the coherence of Albert's position despite its counterintuitive character and the problems it raises by apparently erasing the distinction between natural motions and violent motions (Section 4). The reception of this debate by later authors, such as Blasius of Parma and Luis Coronel, illustrates the importance of this question for late medieval physics (Section 5).

**Keywords** John Buridan · Albert of Saxony · Blasius of Parma · Luis Coronel · Motion

## 3.1 Introduction

The distinction between natural motion and violent motion is a central concept in Aristotelian physics. During the Middle Ages, it remained one of the fundamental notions around which explanations of nature were organised. However, a little-known debate took place in the fourteenth century concerning the scope of this distinction,

S. Roudaut (✉)
CNRS–SPHERE (Paris Cité University), Paris, France
e-mail: sylvain.roudaut@u-pariscite.fr

 
E. Jung et al. (eds.), *The Forgotten Giants: Overlooked Revolutions in Natural Philosophy*, Studies in History and Philosophy of Science 70,
https://doi.org/10.1007/978-3-032-34008-5_3

leading some authors to revise it. More precisely, the debate opposed John Buridan (ca. 1300–1358) and Albert of Saxony (ca. 1320–1390). Both Buridan and Albert of Saxony agreed that the falling of a heavy body, for example, constitutes a case of natural motion. Yet they diverged in explaining how a body can cause an action contrary to its own nature, for instance, how a naturally cold body, such as water, can warm another body, as when one is scalded by hot water.

The context of such a debate is primarily that of the commentaries on Aristotle's *Physics* (although, as we shall see, Albert of Saxony also addresses it in his commentary on *On the Heavens*). In the second book of this work, after explaining the necessity of the three principles—matter, form, and privation—Aristotle undertakes to define what is meant by "nature." Nature, he writes, is the principle of motion and rest in natural things—the notion of motion including, in an Aristotelian framework, not only local motion but also quantitative change and alteration.[1] Yet understanding precisely what this definition entails is far from straightforward, for it requires clarifying what is meant by a "principle of motion".

Within this framework, John Buridan and Albert of Saxony struggle to determine what, in physical bodies, constitutes the main principle of *action*, that is, this particular type of motion by which a body affects other bodies. Their inquiry aims at analysing the internal composition of substances in order to identify this "principle of action" within them. While cases of evidently natural motions—such as fire's natural tendency to heat—pose no major difficulty for them, the problem becomes more complex in the case of motions contrary to a body's own nature. If hot water, as common experience shows, can in turn heat another body, how is one to explain this action given that water is essentially cold by nature? What, in heated water, is truly acting to cause a heat that is contrary to water's own essence? One may of course suppose that it is the *heat* itself that acts, rather than water strictly speaking. But his solution is problematic, as it runs counter to a classical axiom of late scholasticism: *actiones sunt suppositorum* (it is the subjects or substances themselves, strictly speaking, that act, and not their accidents).[2] Yet such actions or motions entail a deeper question: are they to be regarded as natural, or rather as violent—that is, contrary to nature?

In their debate, John Buridan and Albert of Saxony describe this type of action using a terminology characteristic of scholastic thought. In the scholastic framework, actions, in general, can be either immanent or transitive, depending on whether they affect the subject itself (as in the case of intellection, which modifies the subject that performs it) or another subject (for example, a sculptor working on a piece of clay). Transitive actions, in turn, may either conform to the nature of the agent (as when fire ignites paper) or fail to conform to it, as in the case of some water heating another body. A transitive action that does not conform to the nature of the agent, and thus runs contrary to the subject's natural tendency, is nevertheless caused by it. It is precisely this last type of case that John Buridan and Albert of Saxony discuss

[1] Aristotle, *Physics* II.1, 192b13–15; 192b21–23.

[2] See Cross (2008); De Libera (2006).

at length. Their problematic status explains why such actions are not necessarily classified as "violent," as we shall see.

This debate might at first seem to reflect obscure quibbles proper to late scholastic physics. However, its significance becomes clear when considering the difference that separates modern physics from medieval physics. Modern physics, which generally rejects alteration as a type of motion distinct from local motion, relies on the abolition of any distinction between violent and natural motion. If this debate played a role in the deconstruction of the central medieval distinction between violent and natural motion, then tracing its history may be helpful for explaining the emergence of new conceptions of nature typical of modern physics or, at least, a gradual change in what was still considered an 'Aristotelian' understanding of nature.

This study is structured as follows. In the next section, I examine the theoretical background that led John Buridan and Albert of Saxony to question the problematic cases of motion just mentioned (Section 2). After that, Section 3 will show how the positions of Buridan and Albert of Saxony differ from one another. I will then suggest that Albert of Saxony's position, despite seeming more counterintuitive at first glance, presents greater philosophical systematicity (Section 4). Finally, I will briefly address the legacy of this debate among late medieval and early modern authors by examining its reception in the works of Blasius of Parma and at the beginning of the sixteenth century (Section 5).

## 3.2 Historical Background: Self-Motion and the Intrinsic Causes of Natural Motion

Understanding the origins of the debate between John Buridan and Albert of Saxony requires some preliminary historical background on the problem of self-motion or, more precisely, on the idea that the causes of a body's motion may be intrinsic to that body. Aristotelian physics famously asserts the principle *omne quod movetur ab alio movetur* ("everything that is moved is moved by something else"). In the Latin tradition, this principle was commonly interpreted until the end of the thirteenth century to exclude any genuine self-motion of natural bodies.[3] As is well-known, Aristotelian physics relied on the distinction between different types of causes (formal, material, moving or efficient, and final). Applying to the different types of motion, the principle seemed strictly to rule out, in particular, the possibility that a body could be its own efficient cause of motion. This strict rejection of self-motion raised difficulties when confronted with certain natural phenomena, such as the free fall of heavy bodies or the ascent of light ones.[4] Thinkers such as Albert the Great (ca. 1200–1280), Thomas

---

[3] On this principle and its role in explaining the natural motion of elements, see Weisheipl (1985a); Weisheipl (1985b); Pines (1961); Maier (1952), on which this brief summary of thirteenth-century positions draws.

[4] Vucu (2018, 5–6). Besides the studies just mentioned, see also Wallace (1972).

Aquinas (ca. 1225–1274) or Godfrey of Fontaines (ca. 1250–1309) wanted to maintain that, whenever a body is moved, there must exist some distinct mover that causes this motion.

The solution (defended by Albert the Great and Thomas Aquinas among others) that best preserved the thesis of a real distinction between cause and effect in the order of motion consisted in affirming that the efficient cause of a motion—for example, the fall of a heavy body—is, strictly speaking, its generator (i.e., the agent that generated the body). It is essential to have in mind here that the explanations sought by thinkers of this period presupposed the Aristotelian hylomorphic framework, according to which natural bodies are composed of accidents inhering in a substance itself made up of matter and a substantial form. With that in mind, let us explain the reasoning behind this solution. Although this motion, e.g., a heavy body's fall, is due to the nature of this body (e.g., earth), it is not the substantial form of that body as such that acts as the efficient cause of the fall, since this would violate the prohibition of self-motion. The cause must therefore be that which imparts this nature and all its accidents (including heaviness), that is, the generator of the body—in the case of an animal, its parents; more generally, the efficient cause of the body's generation for all beings, including inanimate ones. Admittedly, a factor such as the releasing of an object can also be considered a cause, since it triggers the fall of the object. But the hand that holds an object in the air, when it ceases to hold it and lets it go, does not actively push the object downward. Medieval thinkers appealed to the distinction between an essential cause and an accidental cause in order to express that releasing a heavy object from a certain height, for example, is a cause only in a certain sense, but does not explain the body's active motion downward. Thus, defenders of this position considered the removal of an obstacle—the *removens prohibens* in the technical terminology of the time—only an accidental cause as opposed to the essential (or *per se*) cause represented by the generator of the body.[5] It is important to note that, although the generator acts as the *per se* cause (or mover) of the body's fall, no direct contact with this mover is required during the fall.

At the end of the thirteenth century, several thinkers challenged this view. The context that stimulated the search for new solutions accepting the self-motion of bodies is closely connected to the notion of will and to what may be described as the "voluntarist" current that animates psychology and moral philosophy in this period. The challenge lay in accounting for the nature of the will as a faculty capable of determining or moving itself—following a Christian tradition conveyed above all, though not exclusively, by Augustine—a notion difficult to reconcile with the prohibition of self-motion at the heart of the Aristotelian system.[6] In order to establish the possibility of such a faculty, some authors came to defend, by way of an *a fortiori*

---

[5] See, e.g., Thomas Aquinas, *Sententia libri Metaphysicae*, l. V, lect. 3, n. 13: "Unde removens prohibens dicitur movens per accidens; sive illud accidens sit contrarium, sicut cholera prohibet frigiditatem, unde scamonaea dicitur infrigidare per accidens, non quia causet frigiditatem sed quia tollit impedimentum frigiditatis, quod est ei contrarium, scilicet choleram, sive etiam si non sit contrarium, sicut columna impedit motum lapidis, unde removens columnam dicitur per accidens movere lapidem superpositum alio modo [...]."

[6] See Lang (1992, 63–84).

argument, the possibility of the self-motion of bodies: if even the simplest beings, such as inanimate objects, can in some sense move themselves, then there is nothing surprising in affirming that an immaterial faculty such as the will could do the same.

Thus, Henry of Ghent (ca. 1217–1293), an advocate of this conception of the will, defended the idea that even elementary corporeal substances move themselves. Drawing on Averroes, an important source in this debate in the thirteenth century,[7] Henry affirmed that the heaviness (*gravitas*) of a body functions as the *per se* mover in the body's fall, while what is moved is the body understood as a whole, i.e., the composite of matter and form.[8] Several thinkers after Henry further developed this idea. Indeed, while allowing the mover and the moved to be items present *within* a body, Henry maintained a distinction between them. John Duns Scotus (ca. 1265–1308) radicalised this line of thought by affirming explicitly that a single being could be both mover and moved. Although the details of his argument are too complex to set out here, it should be emphasised that the consequence of Scotus' theory is that the same being can be a mover *per se* and moved *per se*.[9]

The Scotist position was influential in the fourteenth century. John of Jandun played a significant role in its diffusion. Beyond his well-known influence on John Buridan—who pertains to the generation succeeding his at the Parisian Faculty of Arts—the thesis remained associated with Jandun's name in later doxographies, such as that of Nicoletto Vernia and his influential treatise on the motion of heavy and light bodies in the fifteenth century.[10]

In his commentary on Aristotle's *Physics*, Jandun himself acknowledges that this view was not his own invention but derived from a certain "most subtle expositor of Aristotle and his Commentator" among his contemporaries—a possible reference to Scotus, who was active in Paris at the turn of the fourteenth century while Jandun was a student there:

> I nevertheless say, for younger [students], that according to the Commentator it is not impossible for one and the same thing, simultaneously and with respect to the same essential part—namely, to its substantial form—to have essentially (*per se*) both an active power and a passive or receptive power in relation to the same thing according to different respects, though in the same respect this would be truly impossible.
>
> […] For the heavy body, on account of its form—insofar as it is its form—is in virtual act with respect to its motion, and on account of the same form, insofar as it is in matter, it is in receptive potency to that motion.
>
> […] Let later [students] also know, friends of truth more than of fame, that I did not at all find this position from my own invention, but from the teachings of a certain Doctor of

[7] On the crucial role of Averroes on this debate in the Latin world, see Maier (1952); Hassing and Macierowski (1992).

[8] Henry of Ghent (1983, IX, q. 5: 117): "Et est per se movens scilicet forma accidentalis gravis, per se vero motum ipsum corpus gravis compositum ex materia et forma substantiali." For detailed comments, see Vucu (2018, 197–199).

[9] John Duns Scotus (1997, IX, q. 14: 625–673). Cf. King (1994); Effler (1962).

[10] Vernia (1505).

**Table 3.1** Summary of the main positions regarding the prohibition of self-motion

| | 1 | 2a | 2b | 3 |
|---|---|---|---|---|
| | Albert the Great/ Aquinas | Henry of Ghent | Anonymous (mentioned by Vernia) | Duns Scotus/ Jandun |
| *Mover per se* | Generator | Intrinsic quality | Substantial form | Form |
| *Moved per se* | Body | Body | Body | Form |
| *Type of distinction* | Two subjects | Two principles with the same subject | | Ø |

Theology, whom I see among my contemporaries to be one of the most subtle interpreters of the Commentator and of Aristotle […].[11]

Two observations can be made at this point. First, it may be noted that the earliest stages of this debate concerning the possibility of the self-motion of bodies display a careful attention to the various components of a substance involved in motion, as well as to what precisely is moved and in what sense (*per se* vs *per accidens*). For example, Henry of Ghent maintains that a quality moves itself by moving the composite in which it inheres, whereas Duns Scotus defends the view that the same form can move itself. Jandun further specifies that the form moves matter only accidentally, since it primarily moves the whole composite essentially (*per se*). The various ways of combining the possible answers to this problem are catalogued in a fairly systematic manner by Nicoletto Vernia in his treatise, which also reports the views of anonymous authors proposing variants of these positions—for instance, a view close to that of Henry, but maintaining that the mover is the substantial form rather than a quality.[12]

When considering the range of possible positions on this problem—from the strict prohibition of self-motion for a single body, grounded in a real distinction between cause and effect, to its outright rejection—one may identify three main types of positions (the second of which might be further subdivided depending on the exact principle of motion one identifies). These culminate, in the Scotist position, in the affirmation that one and the same form is capable of moving itself (see Table 3.1).

These disputes concerning the best formulation of the self-motion of bodies became relatively less intense by the mid-fourteenth century, at least within Parisian

[11] John of Jandun (1551, VIII, q. 11): 118vbG–119raE [translations are mine]: "Dico tamen propter iuniores, quod secundum Commentatorem, non est inconveniens unum et idem simul et secundum eandem partem essentialem, scilicet secundum formam substantialem suam, habere potentiam activam et potentiam passivam sive receptivam per se eiusdem alio et alio modo, sed eodem modo bene esset impossibile […]. Nam ipsum grave per suam formam, secundum quod est eius forma, est in actu virtuali respectu sui motus, et per eandem formam, secundum quam est in materia erit in potentia receptiva ipsius. […] Sciant etiam Posteriores veritatis plus quam famositatis in animis amici, quod hanc positionem non omnino probavi ex inventione mea, sed ex traditionibus cuiusdam doctoris Theologi, quem inter meos contemporaneos puto unum de subtilibus expositoribus Commentatoris et Aristotelis […]."

[12] On this opinion, see Vernia (1505: 13va); on Jandun and Scotus, see 12va–13rb; on "theologians" holding a view similar to Henry's, see 13rb.

physics. One might reasonably suppose that, as the general thesis of an intrinsic efficient cause of motion in inanimate bodies became increasingly accepted, the fine-grained disputes over its formulation came to be regarded as less necessary, insofar as there was less perceived need to justify it against theories of motion deemed more orthodox.

A second observation is that Parisian natural philosophy in the fourteenth century relied on a particular conception of the structure of natural substances. Although this conception does not directly concern the problem of motion, it is worth recalling here, as it will prove useful for understanding the theoretical presuppositions of the debate and some of the arguments involved. A core thesis shared in this trend, exemplified by John Buridan and Albert of Saxony, was that all components of a substance (its matter, its substantial form, and its accidents) are actual and can, at least supernaturally—in virtue of God's power—exist separately from one another. This view contrasts sharply with earlier conceptions according to which matter is essentially potential and inseparable from a substantial form, just like accidents are inseparable from a subject.[13]

The following claims are thus accepted by Buridan and Albert:

(a) matter and form are actual rather than merely potential.
(b) matter and form can exist separately (though only supernaturally).
(c) real accidents such as whiteness or heat can exist separately from a substance.

These principles have significant implications. Of course, inquiring into the nature of the principle of action in bodies does not necessarily require adhering to these theses; conversely, however, accepting them does entail certain consequences. One may, for instance, investigate what, within substances, truly moves bodies by imagining situations in which these components are separated from one another, and by considering the consequences of such hypothetical scenarios.

This historical background is crucial for understanding the subsequent debate. As we will shortly see, both Buridan and Albert of Saxony accept the idea that the substantial form of a body accounts for its natural motion. Yet this framework does not address the case of transitive actions that do not conform to a body's nature. The famous theory of impetus, defended in particular by Buridan and Albert for explaining the violent motion of projectiles, can be seen as an explanation for cases in which this type of change concerns motion with respect to place, that is, local motion (see Fig. 3.1).

However, no comparable explanatory model existed for motions not conforming to a body's nature that are produced through qualitative change (alteration), such as hot water burning a hand that touches it. It is precisely at this juncture—far less studied than the famous impetus theory—that the debate between Buridan and Albert of Saxony truly begins.

---

[13] For an overview, see Pasnau (2011, 99–199), and the bibliography therein. On the special problem of the separability of accidents, the most complete study to-date remains Bakker (1999).

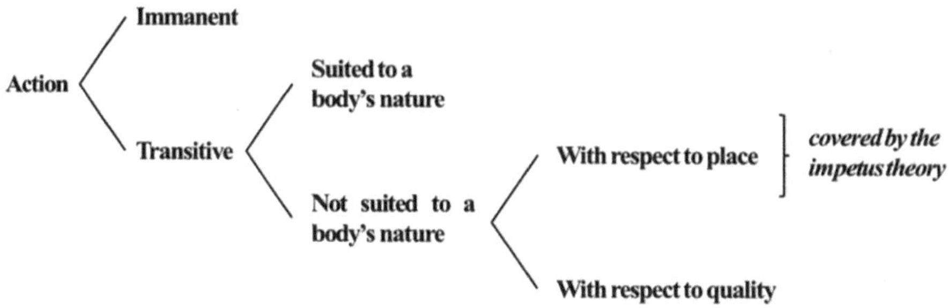


**Fig. 3.1** Classification of types of actions

## 3.3 John Buridan Vs Albert of Saxony on the Cause of Violent (Qualitative) Motion

The major ideas of John Buridan and Albert of Saxony on motion, as well as their acceptance of a theory of natural motion based on an intrinsic cause, have been generally well studied,[14] but the debate we are about to examine has been overlooked by scholars.[15] It is addressed by Buridan in his *Questions on Aristotle's Physics*, and our focus here will be on the final redaction of this work (the so-called *ultima lectura*), composed between 1355 and 1358. This version takes into account Albert's own *Questions on the Physics*, written after 1351. In his own *Questions on the Physics*, however, Albert does not confront the problem head-on but addresses it only partially, by indicating how to evaluate propositions concerning the intrinsic principles of natural transmutation in light of the semantics he adopts—namely, the theory of supposition.[16] It is rather in his *Questions on Aristotle's On the Heavens*, likewise dating from his Parisian teaching, that Albert engages more directly with the issue, responding to the position defended by Buridan. Although the precise dates of composition of these works remain unknown, it is clear, as we shall see, that they are in dialogue with one another and that both authors were fully aware of the positions they were criticising, even if they refrained from naming each other explicitly (as was often the case in this period).

### 3.3.1 *John Buridan*

John Buridan's position on the cause of natural motion, widely shared by leading Parisian masters such as Albert of Saxony, Nicole Oresme, and Marsilius of Inghen,

[14] See Sarnowsky (1989) for a detailed study of Albert of Saxony's theory of motion and its relation to John Buridan's; see also Thijssen (2009). See Sylla (1993) for an analysis of the intrinsic causes of natural motion.

[15] However, see Zambiasi (forthcoming) for the later reception of this debate in Vienna.

[16] Albert of Saxony (1999, I, q. 19: 247–258).

rests on a key distinction.[17] In the *Physics* (III.2, 202a9–12), Aristotle had already defined form as the principle of change; however, in the fourteenth century, more refined distinctions—such as that between a "principal principle" and an "instrumental principle," characteristic of late Scholasticism—required a more precise formulation of the issue. Buridan accepts the view that the efficient cause of a heavy body's fall is its substantial form and its heaviness (which is an accidental quality of the body). While the former represents the *principal* (efficient) cause or principle of natural local motion, the latter represents its *instrumental* cause.[18] The substantial form represents the principal cause in the sense that it generates heaviness, which in turn effectively produces the downward motion of a heavy body. These causes, which are intrinsic to a body, are the *per se* mover of the body's fall.[19] In light of the classification of positions seen above, Buridan's position, like Albert's as we will see, is more akin to the second, identifying as principles of motion both the substantial form of a body (2b) and an accident (2a) in the natural case—although Buridan sometimes seems to admit that a form separated from all matter could also (in a supernatural hypothesis) move itself directly.

This position presupposes a principle of causal contact: for an effect to exist, it must remain in continuous contact with its cause (*causa et effectus sunt simul*, according to a standard scholastic axiom). The causal efficacy of substantial forms is therefore not limited according to Buridan to the initial generation of a substance but is constantly present in all its natural motions.

Importantly, Buridan applies this same framework to processes of alteration. The substantial form of some fire, for instance, is itself the principal cause of the heating activity of this fire; similarly, the substantial form of water is the principal cause of its coldness.[20] This understanding of the intrinsic causal structure involved in all natural motions leads Buridan to think that actions that are contrary to a body's nature must have a different cause, a point he makes explicit in his *Questions on Aristotle's Physics* (II, q. 5):

> It should be noted that in every natural substance having a substantial form in matter, there are changes or even rests that are suitable to its nature, as well as to its health and perfection, but there are also often [such dispositions] that are not suitable to its nature [...]. Thus, cooling and resting in coldness are suitable to the form of water; heating and resting in heat, on the other hand, are contrary to its nature.
>
> From these observations, it seems to me that it must first be said that the substantial form is not the active principle by itself of changes and rests contrary to its nature, for it rather resists and inclines to the opposite state. [...] But it must be said that the substantial form is the active principle of changes and rests suitable to its nature. And this is clearly seen, first of all, in the soul of animated beings. [...] The soul does not act only with heat; it also produces this heat and preserves it.[21]

---

[17] See Sylla (1993) for an overview.

[18] Buridan (2015, II, q. 4: 267, ll. 7–11).

[19] Buridan (2015, II, q. 4: 267, ll. 19–21; 268, ll. 5–13).

[20] Buridan (2015, II, q. 4: 267, ll. 12–13).

[21] Buridan (2015, II, q. 5: 277–278, ll. 15–22): "Tunc ultra notandum est quod in omni substantia naturali habente formam substantialem in materia inveniuntur aliquando transmutationes vel etiam

Buridan repeats this statement when discussing the priority of substantial forms over accidents in the causal hierarchy, thereby justifying the designation of substantial forms as "principal" causes or principles:

> Substantial forms are the primarily active principles of changes and rests that are suitable to its nature, and not the accidents that are joined to it because, when two agents contribute to a certain effect, the one that is the most active and brings the other back to its nature (*ad convenientiam*), if [the other] had departed from it, directing it in acting, is the principal one, like the soul with respect to heat, or water with respect to coldness.[22]

In other words, substantial forms are the most fundamental agents of natural change: they are responsible for producing the qualities proper to their respective substances, for restoring them when they deviate from what is appropriate, and for directing their activity in accordance with the nature of the body. Yet Buridan also insists that some motions do not arise from the nature of the substance itself. As he observes regarding vapour:

> This is why it must be clearly conceded that vapour does not move upward by its nature; it does not move upward but, rather, it is moved upward by an accident it acquired extrinsically and which does not suit its nature.[23]

According to Buridan, while substantial forms are the primary agents of natural changes suited to a substance's nature, motions contrary to it must be traced to accidents acquired extrinsically. Trying to reconstruct the principles supporting Buridan's view, we can note that it rests on three related main commitments. First, as noted above, Buridan upholds a causal contact principle in the case of motion. This principle might not be universally true for all types of causes, since the term 'cause' may be taken in different senses, but it holds with respect to things coming to being, i.e., things in the process of being made, including motion as the actualisation of a new thing or state.[24] This idea appears to underlie Buridan's conception of substantial form and qualities as the immediate intrinsic source of actions performed by a body.

---

quietes convenientes illi formae et ad eius salutem et perfectionem, et inveniuntur saepe, et aliquando disconvenientes ei [...]. Et sic frigefactio et quies in frigiditate sunt convenientes formae aquae, calefactio autem et quies in caliditate sunt eidem disconvenientes. Ex notatis videtur mihi dicendum primo quod forma substantialis non est principium per se activum ad mutationes et quietes sibi disconvenientes, quia potius resisteret et inclinaret ad oppositum. [...] Sed dicendum est quod forma substantialis est principium activum ad mutationes et quietes sibi convenientes. Et hoc primo manifestum est de anima in animatis. [...] Item anima non solum agit cum calore, immo etiam agit ipsum calorem et conservat."

[22] Buridan (2015, II, q. 5: 279–280, ll. 25–4): "Istis visis concludendumest tertio quod formae substantiales sunt principia principaliter activa ad mutationes et quietes sibi convenientes et non accidentia eis coniuncta quia: duorum agentium ad aliquem effectum concurrentium illud est principale respectu alterius quod est activum ipsius et reductivum ipsius ad convenientiam, si fuerit disconveniens, et directivum ipsius in agendo; sed sic se habet anima respectu caloris, ut dictum est, sic etiam se habet aqua respectu frigiditatis."

[23] Buridan (2015, II, q. 5: 281, ll. 11–14): "Unde concedendum esset plane quod vapor non movet se sursum per suam naturam, immo non movet se sursum, sed accidens sibi ab extrinseco acquisitum et sibi disconveniens movet ipsum sursum." Cf. II, q. 8: 304, ll. 1–16.

[24] See Buridan (2015, II, q. 8: 303, ll. 6–14).

Second, his framework presupposes that either a substantial form or an accident may operate as the efficient cause of a physical change and, in particular, of a transitive action. Finally, Buridan affirms a *duality* of principal principles of actions. Indeed, substantial forms operate as the principal agent when the change accords with the substance's nature—for example, the soul producing and maintaining heat in a living body—whereas qualities (acquired extrinsically) are the principal causes of changes contrary to a body's nature, such as the upward motion of vapour or the heating effect of hot water. The second of these commitments, as we will now see, is contested by Albert of Saxony, leading him consequently to also reject the third one.

### *3.3.2 Albert of Saxony*

Albert's natural philosophy shares important features with that of his contemporary Buridan. But whereas Albert adopts, like Buridan, many doctrines typical of Parisian physics concerning motion—such as the impetus theory and the so-called "Bradwardine rule" of motion—he diverges sharply from him on the issue of non-standard cases of transitive causality. Albert's critique is most clearly expressed in his *Questions on Aristotle's On the Heavens*:

> Let us briefly see that this opinion is not entirely true. First of all, because it assumes that certain accidents are the main active principles of certain effects, such as, for example, heat in hot water being the main principle of heating, and not the substantial form of water. Similarly, they say that lightness is the main active principle of the upward motion of vapour and not the substantial form. But this is incorrect. For if heat alone, without the substantial form of this hot body, could produce heat, then it could also produce fire. However, this is impossible. Indeed, what produces would then be less noble than what is produced. What produces, in fact, is an accident, for example heat, and what is produced is a [substantial] form, for example fire. Now, a substance is more noble than an accident. The falsity of the consequent is clear in light of this common rule, namely that nothing acts beyond its own degree or species.[25]

According to Albert, it is false to claim that accidents may act as the principal causes or principles of actions. Albert's refusal to grant accidents such a causal role rests on a deep metaphysical conviction, which can be described as a principle of 'causal superiority': a cause must always be equal to or more perfect than its effect. Since accidents are ontologically inferior to substances, they cannot generate effects of a

[25] Albert of Saxony (2008, III–IV, q. 9: 502–503, ll. 68–82): "Breviter videtur quod illa opinio non sit omnino vera. Primo quidem propter hoc quod supponit aliquod accidens esse principale principium activum alicuius effectus, sicut verbi gratia quod caliditas in aqua calefacta sit principale principium calefactionis, et non forma substantialis aquae; et similiter ascensus ipsius vaporis dicunt principale principium activum esse levitatem, et non formam substantialem. Sed illud non valet. Nam, si caliditas se sola, sine forma substantiali ipsius calidi, posset producere caliditatem, etiam sic posset producere ignem; modo hoc est impossibile, nam tunc producens accidens, puta caliditas, et productum esset forma, puta ignis; substantia est nobilior accidente. Falsitas consequentis patet per istam communem regulam quod nihil agit ultra gradum proprium suae speciei."

higher ontological rank, such as substantial forms.[26] Let us suppose for instance that a sheet of paper is ignited by means of an iron bar—naturally cold—heated by an external source of heat. If we assume, as Buridan does, that the heat present in the bar ignited the paper, then we must suppose that an accident was able to produce a substance, which is metaphysically impossible.

Thus, Albert defends the view that substantial forms are *always* the principal principles of actions in bodies. He defends this view against an obvious objection:

> The argument in favour of this opinion is not convincing, namely that if vapour were to rise by its substantial form, having lightness, and descended by that same form, having heaviness, a single simple form would produce distinct motions. This is entirely true, and I see no problem with it, when this happens through different instruments. Doesn't the intellective soul, which is an absolutely simple form, have entirely diverse operations through different instruments? It sees with certain instruments, touches with others, and hears with others. Similarly, I say in the present case that there is no problem with the substantial form of water heating through heat and cooling through coldness, producing upward motion through lightness, and producing downward motion through heaviness. [...] These movements arise from the substantial forms of heavy and light bodies as principal agents, although they arise from lightness and heaviness as instrumental agents. [...] *For such motions are natural* [...] they must arise from the nature of heavy and light bodies. And since there are [in these bodies] only two natures, namely matter and form, and there is no activity in matter, they can only arise from the substantial form.[27] [emphasis added]

As can be seen, Albert is not impressed by the objection according to which one single form (like that of an element) can produce only one type of motion, since one single substantial form may cause different actions through different "instruments," i.e., qualitative dispositions.

Albert's account of transitive action contrary to a body's nature can be reconstructed as an opposition to John Buridan's framework, although he shares with him some theoretical principles. On the one hand, Albert also accepts the validity of the principle of causal contact in cases involving an efficient causal relationship between two things, when the first produces or sustains the second in being.[28] He also accepts

[26] On this principle, see Roudaut (2023).

[27] Albert of Saxony (2008, III–IV, q. 9: 503–504, ll. 82–14): "Nec valet ratio istius opinionis, scilicet quod, si vapor ascenderet per suam formam substantialem habens levitatem et postea descenderet per eandem habens gravitatem, eadem forma simplex esset productiva distinctorum motuum. Certe verum est, nec hoc reputo inconveniens, quando hoc fit mediantibus diversis instrumentis. Nonne anima intellectiva, quae est forma valde simplex, valde diversas habet operationes mediantibus diversis instrumentis: unde per aliqua instrumenta videt, per aliqua tangit, et per aliqua audit, etc. Sic dico in proposito non esse inconveniens eandem formam substantialem aquae mediante caliditate calefacere, et mediante frigiditate frigefacere, et mediante levitate producere motum sursum, et mediante gravitate producere motum. [...] Dicti motus proveniunt a formis substantialibus gravium et levium tamquam ab agentibus principalibus, licet a gravitate et levitate proveniant tamquam ab agentibus instrumentalibus. [...] Nam tales motus sunt naturales sicut supponitur in conclusionibus; ideo oportet quod proveniant active a natura gravium et levium tamquam ab agente principali. Et, cum in gravi et similiter in levi non sint nisi duae naturae, scilicet materia et forma, oportet quod dicti motus proveniant active vel a materia vel a forma, et hoc substantiali. Sed non a materia, cum illa nullius sit activitatis; reliquitur igitur quod a forma."

[28] Albert of Saxony (1999, II, q. 10: 382–383, ll. 76–97).

that the efficient cause of a natural action or motion is intrinsic to the body acting or in motion.

On the other hand, Albert diverges sharply from Buridan on the question of principal principles, rejecting the 'dual' solution advocated by Buridan. In Albert's view, there can be only one kind of principal cause or principle in natural changes—namely, the substantial form. Accidents, such as heat or lightness, cannot by themselves serve as sufficient causes of any effect; they operate "instrumentally," subordinate to the activity of the underlying substantial form. According to Albert, even if the heaviness of a body were supernaturally detached from its matter and substantial form, it would not move downward, for such a motion can only occur through a principal principle of motion, i.e., a substantial form.[29]

This difference is noteworthy, and the strength of Albert's solution lies here in his use of the notion of instruments. As we have seen, Buridan also admitted that the qualities proper to a body's nature could be called "instrumental causes" subordinated to the body's substantial form—a terminology that had been common on this subject since at least Thomas Aquinas.[30] Albert, however, reverses this terminology, employing it to conceive the causality of natural substantial forms (even in inanimate bodies) on the model of rational human agency, in which the principal cause (the intellect) makes use of various instruments to produce opposite actions, including those contrary to a body's nature. Here, Albert is probably exposing himself to the kind of criticism later advanced by Descartes, who would mock the Scholastics' appeal to "little souls" present in inanimate forms.

## 3.4 Comparative Analysis

The debate between John Buridan and Albert of Saxony on the nature of principles in physical actions reveals two markedly different strategies for reconciling an Aristotelian physical terminology with the conceptual constraints of fourteenth-century natural philosophy. In this section, let us attempt to set out systematically the implications of their respective positions.

### *3.4.1 Natural Vs Violent Motion*

At first sight, Buridan's position appears more faithful to the Aristotelian distinction between natural and violent motion. Like Aristotle, Buridan defines violent motion as that against which a body resists, insofar as it is contrary to its own natural inclination.[31] Buridan can thus fully preserve Aristotle's taxonomy of motions and

[29] Albert of Saxony (2008, III–IV, q. 10: 508, ll. 98–00); cf. Buridan (2015, II, q. 5: 282, ll. 6–13).

[30] On instrumental causality in the Middle Ages, see Ehret (2021); Solère (2019).

[31] Buridan (2015, I, q. 21: 215, ll. 21–25).

explain the forced behaviour of bodies through accidental dispositions that oppose the underlying nature of the substance.

Albert, by contrast, proposes a much more unorthodox and, in certain respects, radical solution. For him, all motions—whether seemingly natural or violent—must ultimately be traced back to the substantial form as the principal cause. The accidents of lightness, heaviness, heat, or coldness, although acknowledged, are reduced to the status of instrumental causes. This move significantly simplifies the causal framework but this simplification comes at a cost: it blurs the very boundary between natural and violent motions around which Aristotelian physics is articulated. Indeed, since transitive actions contrary to nature are due to the substantial form, and since the latter is a nature (as the principal cause of motion), it follows from this position that such motions are *natural*.

In fact, Albert also accepts—like Buridan—the common definition of violent motion as a motion contrary to a body's inclination.[32] This definition is of course expected in a context of commenting Aristotle's *Physics*. In practice, however, Albert tends rather to employ a different characterisation of violent motion, namely, that a violent motion is one caused by an extrinsic mover.[33] Given Albert's acceptance of a theory of natural motion based on intrinsic efficient causes, this characterisation, also taken from Aristotle,[34] has somewhat different implications. Violent motions certainly exist in the case of local motion; however, for cases in which actions are due to a quality inhering in the subject, it follows from this characterisation that transitive actions contrary to nature are not necessarily violent (as evidenced in the last passage cited in the previous section). For Albert, the only genuinely 'violent' phenomena with respect to qualities is the process during which a body temporarily acquires an accident wholly alien to its nature—for instance, when an iron bar is externally heated. What the heated iron bar subsequently *does* to other bodies, in turn, is natural.

Some cases might be difficult to determine for Albert. For example, in his *Questions on Aristotle's On the Heavens*, he mentions dust particles floating in the air, visible in the light from a window. If one chooses to explain this by saying that they are heavy but too small to overcome the resistance of the air to their fall, then one must say that their rest is *violent*. If, however, one admits that they rise thanks to the heat and lightness induced in them by sunlight, then the situation is different.[35] As a matter of fact, Albert also notes in his *Questions on the Physics* that violent and natural motions can be "of the same species," in the sense that they can be indiscernible regarding their end terms and trajectory.[36] In short, Albert perhaps does not depart so very far from the Aristotelian definition of violent motion; yet the way he adapts it to his own explanation of motion considerably reduces the number of

[32] Albert of Saxony (1999, II, q. 1: 291, ll. 27–29).

[33] Albert of Saxony (1999, III, q. 6: 513, ll. 33–39); Albert of Saxony (2008, I, q. 1: 13, ll. 73–75).

[34] Aristotle, *Nicomachean Ethics* III.1, 1110b15–16.

[35] Albert of Saxony (2008, III–IV, q. 1: 454, ll. 11–22).

[36] Albert of Saxony (1999, V, q. 6: 806–807, ll. 24–35).

motions that can be qualified as violent, by excluding many cases where such motions result from qualities inhering in a subject.

### 3.4.2 *Metaphysical Principles of Causation*

A major point of divergence between the two thinkers concerns the principle of causal superiority, according to which a cause must be equal to or more noble than its effect. Both authors agree on its validity, but they do not draw the same conclusions from it. Contrary to what Albert's critique suggests, Buridan also denies that an accident can ever produce a substance. Thus, his framework forces him into an awkward position: because certain substantial changes—those triggered by accidents contrary to a substance's nature—cannot be accounted for by natural principles alone, Buridan must posit periodic divine interventions in the course of nature. He openly acknowledges this recourse, citing authorities such as Plato, Avicenna, and Themistius to justify the need for a transcendent "Giver of Forms":

> And it seems to me that this was the main reason why Plato posited separate Ideas, and why Avicenna and Themistius concluded that the Giver of forms is a substance separate from magnitude—that is, not extended, nor occupying any determinate place in the world, but present and without distance to the whole world and to each part of the world, namely God, who, being thus present to the whole world without any distance, can act upon it wholly and upon any part of it.[37]

This appeal to divine causality is far from marginal: it becomes necessary whenever accidental motions lead to substantial transformations, as in certain cases of violent motion or in the generation of living beings—especially in animal reproduction, where the accidental heat of the male seed is thought to initiate substantial change in the female matter. Returning to our example of the iron bar, the strangeness of Buridan's position becomes apparent: if I burn myself by touching this bar, this action requires nothing other than the operation of the accident (heat) producing this affection in my flesh. If, however, I bring a sheet of paper near the bar and it is set alight by it, God must intervene (Buridan does not specify whether such intervention is punctual or general), without which the sheet of paper could not catch fire (that is, be substantially transformed into fire). Thus, physical processes that are, all things considered, quite common in nature require, for Buridan, a supernatural intervention in order to be explained.

Albert's position avoids this theological complication by consistently applying the principle of causal superiority: substantial forms alone are capable of producing other substantial forms. In this respect, Albert's system achieves greater theoretical

---

[37] Buridan (2015, II, q. 5: 276, ll. 20–26): "Et videtur mihi quod ista fuit maxima ratio propter quam Plato posuit ideas separatas et propter quam Avicenna et Themistius concluserant datorem formarum esse substantiam separatam a magnitudine, id est non extensam, nec sibi aliquem situm in mundo determinantem, sed toti mundo et cuilibet parti mundi assistentem praesentialiter et sine distantia, scilicet Deum, qui sic toti mundo praesens et indistans potest agere in ipsum totum et in quamlibet partem eius."

simplicity. With respect to the principle of causal superiority, however, it is not without its own challenges. While it neatly eliminates the need for divine intervention in natural processes, it leaves unresolved (as far as I can see) the broader problem of how a less noble substance can give rise to a more noble one—a difficulty that becomes especially pressing in cases of artificial incubation, for instance.[38]

### 3.4.3 Semantics

From a semantic perspective, Albert's approach also seems to enjoy a certain advantage. Buridan and Albert share a broadly similar conception of the semantics of substantial terms, analysing notions such as 'nature,' 'cause,' and 'principle' in parallel ways. Buridan writes: "Nature, indeed, is the first cause, that is, the principal cause, and the first principle, that is, the main principle, of motion, etc."[39] Ontologically speaking, a nature is a substance, i.e., either matter or substantial form.[40]

Albert presents the same analysis of the term 'nature' and also equates it, following Aristotle's definition, with the (principal) principle of motion and rest.[41] But in Albert's framework, these terms always retain a stable reference to *something substantial*.[42] Although the term 'nature' pertains to an accidental category because it denotes something *ad aliquid*, indicating a reference to motion or rest, it always refers, according to Albert, to a substance (or "supposits for" it, following the theory of supposition of this time).[43] The passive (principal) principle of motion is matter, which is a substance, while the active (principal) principle of motion is the substantial form. 'Nature,' 'principle,' and 'cause' are effectively synonymous and interchangeable as they refer to the same type of thing.

Buridan, by contrast, faces a semantic tension: because his system allows for accidents to function as principal principles in certain contexts, he must explain why the referent of 'principle' sometimes shifts away from its usual signification or, equivalently, why 'nature' cannot always be substituted for 'substance' in certain propositions. The expression 'principal principle' sometimes refers to a substantial form, that is, a 'nature' in the strict sense, but it may also refer to an accident in other contexts.[44] This complication highlights how metaphysical commitments in late medieval physics were often intertwined with fine-grained semantic analyses.

---

[38] This problem appears to have motivated Buridan's solution; see Buridan (2015, II, q. 5: 275, ll. 1–13).

[39] Buridan (2015, II, q. 4: 267, ll. 16–18): "Natura enim est causa prima, id est principalis, et principium primum, id est principale, movendi etc."

[40] Buridan (2015, II, q. 4: 265, ll. 20–21; 267, ll. 14–15).

[41] Albert of Saxony (2008, II, q. 5: 336, ll. 30–42).

[42] Albert of Saxony (2008, II, q. 5: 337, ll. 53–54).

[43] Albert of Saxony (2008, II, q. 5: 335, ll. 99–8). Cf. Buridan (2015, II, q. 4: 265–266, ll. 24–3).

[44] Buridan (2015, II, q. 2: 253, ll. 1–2. Cf. II, q. 5: 282–283, ll. 31–3).

In light of this comparison, one might ask whether the origin of the dispute is simply Albert's choice to consider certain actions—such as those of vapour or hot water—as natural rather than violent. This could be seen as question-begging, yet it compels him to adopt his particular position given his philosophical commitments. At any rate, his stance appears to threaten the structure of Aristotelian physics more directly, by problematising the distinction between natural and violent motion, and probably also the distinction between rational power (capable of contrary actions) and irrational power (capable of only a single type of action), the latter normally corresponding to the motion of the elements.[45] From an external and modern perspective, Albert's position may appear simpler and more philosophically coherent. In its own time, however, Buridan's position likely appeared more faithful to Aristotle's physics, and therefore more appealing, probably explaining why it seemed to have dominated in later doxographies of the debate.

## 3.5 Later Reception of the Debate

Considerable research remains to be undertaken regarding the fortune of this debate. In light of the current state of research, it can nevertheless be affirmed that it exerted an influence and generated a series of reinterpretations extending well beyond the fourteenth century. Already in the final decades of that century, some thinkers explore the implications of the Buridan–Albert controversy for broader issues in natural philosophy.

A particularly interesting echo of this debate appears in the work of Blasius of Parma (ca. 1350–1416), a philosopher, physician, and mathematician active in the intellectual circles of late fourteenth- and early fifteenth-century northern Italy.[46] Of particular relevance here is his *Questions on Aristotle's On the Heavens*, composed around 1379–1382 and inspired by Albert's own questions, which Blasius lectured on. In these questions, Blasius develops a position that reinterprets—and in some respects radicalises—Albert's views on transitive motion and their implications.[47] It is in Book IV, where Albert himself had advanced his theory, that Blasius addresses this issue. Rejecting Buridan's position, Blasius notes that it enjoys considerable acceptance among his contemporaries and describes it as "probable and held by many" (*licet sit probabilis et multi teneant eam*). His argument rests on two underlying assumptions. First, if a quality is capable of exercising causal efficacy upon a subject, then it must also be capable of producing a substantial form, since disposing

[45] Aristotle, *Metaphysics* Θ.2, 1046b4–7.

[46] For documentation, see Biard and Robert (2019); Lagerlund et al. (2025), and the bibliography therein.

[47] See Federici-Vescovini (1991, 240–246).

a subject—or matter—is identical to inducing a form within it. Second, like Albert, Blasius maintains that no quality can in fact cause a substantial form[48]:

> It should not be said, therefore, that in the upward motion of vapour, this motion is produced by the induced lightness as its principal [principle] [...]. For this reason, heaviness and lightness do not actively contribute to the motion of heavy and light bodies. This is evident, since then it will be said that natural motion is produced by a quality rather than by the intrinsic principal principle, namely the substantial form. It should be said, however, that these qualities contribute instrumentally [...].[49]

According to Blasius too, qualities contribute only instrumentally—and not "actively," that is, not in a principal way—to the production of motion. Compared to Albert, however, Blasius endorses a slightly different position that could be roughly described as more "Thomistic" regarding the relation between substances and accidents. He explicitly rejects, for example, the idea that a substantial form could act apart from its accidents, and argues in particular, against Albert of Saxony, that the substantial form of a heavy body, were it separated from matter and located in a celestial region, would not fall down, since it requires its accidents and inherence in matter in order to do so.[50]

In any case—and probably not coincidentally—Blasius appears even more inclined than Albert to overcome the opposition between natural and violent motion. In the same work, he proposes the striking claim that every local motion is natural:

> First proposition: Every generable and corruptible thing is natural [...]. Second proposition: Every local motion is natural. This is proved, because local motion is identical either with the mobile [thing] itself, or with a disposition inhering in it. If it is the mobile thing, and all mobile things are generable and corruptible, as the first proposition says, then it follows that motion is natural. If it is a disposition inhering in it, the same conclusion follows. Third proposition: Every violent motion is natural. This is proved: Every violent motion is something generable and corruptible and every such thing, as the first proposition says, is natural, from which it is concluded that every violent motion is natural. Fourth proposition: The motion by which a heavy [thing] is moved violently is natural. This is clear from what precedes, if the first propositions are admitted. Fifth proposition: Violent motion is not natural motion. This proposition is evident upon inspection: there is no logical way in which the terms on the side of the predicate can be said to appellate their form and nature (*rationem*). Sixth proposition. This one, I argue, must be noted: this motion is natural; therefore, this motion is a natural motion.[51]

[48] Blasius of Parma, *Questiones de caelo*, IV, q. 2: 72r. On this topic, though based on different works by Blasius, see Biard (2022: 65–77).

[49] Blasius of Parma, *Questiones de caelo*, IV, q. 2: 72r: "Non est igitur dicendum quod in motu vaporis sursum ille motus fiat a levitate inducta et hoc principaliter [...]. Ideo gravitas et levitas non concurrunt active ad motum gravium et levium. Patet, quia tunc dicetur quod motus naturalis fieret a qualitate et non a principio principali intrinseco ut a forma substantiali. Dicatur tamen quod istae qualitates instrumentaliter concurrunt [...]."

[50] Blasius of Parma, *Questiones de caelo*, IV, q. 3: 72v.

[51] Blasius of Parma, *Quaestiones de caelo*, I, q. 4: 39v: "Prima propositio: omnis res generabilis et corruptibilis est naturalis [...]. Secunda propositio: omnis motus localis est naturalis. Probatur quia vel motus localis est ipsum mobile vel dispositio sibi inhaerens. Si est ipsum mobile, et omnia mobilia sunt generabilia et corruptibilia, ut dicit prima propositio, sequitur quod motus est naturalis. Si sit dispositio sibi inhaerens, hoc idem sequitur. Tertia propositio: omnis motus

According to Blasius, it is possible to maintain, from a logical point of view, that there is some meaning in speaking of 'violent motion'; yet, from an ontological standpoint, there is no real difference between violent and natural motion. Every motion ultimately refers to a natural thing to which it is reducible, the substantial form being always the intrinsic (principal) active cause in natural beings.

Let us note, importantly, that Blasius will adopt a different approach in his later *Questions on the Physics*. There, he will argue that every natural body seeks preservation and rest in its proper place and, consequently, that no body within the sublunary sphere naturally desires local motion, its natural inclination being toward rest. From this, in a manner diametrically opposed to the position expressed in *Questions on Aristotle's On the Heavens*, Blasius will conclude that all local motion in the sublunary realm must be violent, since it runs contrary to a body's natural tendency.[52]

In other words, there are clearly different versions of Blasius's theory of motion, namely some in which he tends to present all motions as violent, and others where all motions are described as natural. But these versions converge on a fundamental attempt to neutralise the sharp Aristotelian distinction between violent and natural motions. In particular, in his earlier *Questions on Aristotle's On the Heavens*, directly influenced by Albert's peculiar theory, he favours a world picture in which violent motions are absorbed into the broader class of natural ones.

One could certainly speculate here on the role played by Blasius in the dissemination of such ideas, given that he also played a role in the spread of the so-called 'mechanical tradition' or *scientia de ponderibus* (science of weights) and was widely read, including by figures such as Leonardo da Vinci, up to the sixteenth century. The science of weights, or mechanics, is indeed recognised as having been a decisive factor in the gradual erosion of the opposition between natural and violent motion at the dawn of the early modern period.

Yet the debate did not only find echo in the work of Blasius. Traces of it are found in late scholasticism. Quodlibetal questions devoted to this controversy were treated in Central European universities in the fifteenth century, where the works of mid-fourteenth-century Parisian thinkers such as Buridan and Albert were widely disseminated.[53] Another notable case from a later period is that of the Castilian philosopher Luis Coronel in the early sixteenth century, who was active in Paris and belonged to

violentus est naturalis. Probatur. Omnis motus violentus est res generabilis et corruptibilis, et omnis talis est naturalis ut dicit prima propositio, quare concluditur quod omnis motus violentus est naturalis. Quarta propositio: motus quo grave violenter movetur est naturalis. Patet ex praemissa cum admittantur primae propositiones. Quinta propositio: motus violentus non est motus naturalis. Ista propositio patet inspectu: nulla pars logice per quam dicimus terminos a parte praedicati potest appellare suam formam et rationem. Sexta propositio: ista arguo notandum iste motus est naturalis, igitur iste motus est motus naturalis."

52 Blasius of Parma, *Quaestiones super libros Physicorum (secunda redactio)*, II, q. 2, in Lagerlund, Roudaut, and Åkerlund (2025, 196, n. 70): "Et sequitur consequenter quod omnis motus localis ad intra circulum lunae est violentus. Patet postquam est contra inclinationem corporis naturalis." On this, see Åkerlund et al. (2025).

53 Zambiasi (forthcoming).

the intellectual circle of John Mair.[54] In his *Physical Investigations*, Coronel explicitly discusses the opposition between Buridan and Albert, reconstructing their arguments in detail.[55] While he appears to grant a slight advantage to Buridan's position, he refrains from offering a definitive resolution, acknowledging the complexity of this ongoing question.[56]

John Mair himself, for his part, acknowledging the difficulty of the problem in his commentary on the *Physics* (II, q. 4), mentions alongside natural motion and violent motion a third type of motion, which he describes as "neutral." A somewhat later Hispanic author, Diego of Astudillo, also analyzes this debate and rightly notes that Albert's particular position begs the question by assuming that a motion such as the heating of water is a natural motion, whereas for him it is clearly a violent motion.[57] There is no doubt that many other echoes and continuations of this debate remain to be explored.

## 3.6 Conclusion

The debate over the active principles of motion in fourteenth-century natural philosophy between John Buridan and Albert of Saxony highlights both the continuity and the tensions within the Aristotelian framework inherited by late medieval thinkers. Let us look back at the starting point of this story. In the fourteenth century, reflections on the causality at work in natural motions, influenced by debates on the possibility of self-motion, focused on identifying the exact components responsible for the various types of actions of which bodies are capable. In the context of commentaries on Aristotle's physical works—works that present many ambiguities and obscurities—authors such as Buridan and Albert, whose nominalist approach makes their treatment of physical problems fundamentally linguistic, were led to seek as precise a semantics as possible for terms such as 'nature,' 'principle,' and 'cause.' Theories of internal causes of motion defended at the time complicated the issue, for it appeared that most kinds of actions contrary to the nature of bodies—actions normally falling under the class of "violent motions"—were due to internal causes (qualities inhering in the substance), something that should normally define natural motion.

Taking a step back from the debate, one can see that it emerged, out of these premises, from the difficulty of fully reconciling this new way of thinking about the motion of bodies and its causes with certain axioms held to be true: (1) the prohibition of self-motion, (2) the principle of causal superiority, and (3) the principle of causal contact. Buridan and Albert are inclined to admit the third in the case of motions and actions, even if this means amending to some extent the first principle. As we

[54] For information on this author, see Roudaut (2025) and the bibliography therein.

[55] Luis Coronel (1530, I: 16vb–19ra).

[56] Coronel (1530, 17vb): "Aliam positionem Buridani probabiliorem reputamus ut pote que absque minore difficultate et paucioribus inconvenientibus sustineri potest [...]."

[57] Diego of Astudillo (1532, II, q. 5: 50rb).

have seen, they both also accept the second principle, but diverge on the best way to circumvent the difficulties it poses in the case of motions contrary to a body's nature. Buridan's dual solution preserves Aristotelian distinctions at the price of theological supplementation; Albert's own solution achieves logical coherence but erodes the conceptual barriers between kinds of motion. From Albert's perspective, insofar as these motions are primarily due to the substantial form, which is a nature, they are entirely natural.

Somewhat surprisingly, one finds relatively few arguments in this debate pointing to the ruinous consequences of Albert's theory for the central Aristotelian distinction between violent and natural motion. From the perspective of the historian of science and philosophy, it is nonetheless evident that such a theory had important implications for the consistence of the scholastic–Aristotelian framework that, at least terminologically, still dominated the period. The fact that this debate is mentioned by authors such as Blasius of Parma or Luis Coronel—who developed profoundly new ideas on motion—is significant. Only further studies will determine to what extent this debate may have nourished a 'silent' or 'forgotten' revolution within late medieval natural philosophy, paving the way for a new conception of nature, but it is clear, in any case, that it contributed to questioning views on motion that at the time seemed entirely unchallenged.

## References

### *Primary sources*

Albert of Saxony. 1999. Quaestiones in Aristotelis Physicam, edited by Benoît Patar, 3 vols. Louvain-la-Neuve/Leuven: Éditions de l'Institut Supérieur de Philosophie/Peeters.
Albert of Saxony. 2008. Quaestiones in Aristotelis De caelo, edited by Benoît Patar. Louvain-la-Neuve/Leuven: Éditions de l'Institut Supérieur de Philosophie/Peeters.
Blasius of Parma. *Quaestiones de caelo*. Ms. Rome, Biblioteca Angelica, 592 II, 36r–75v.
Buridan, John. 2015. *Quaestiones super octo libros Physicorum Aristotelis (secundum ultimam lecturam). Libri I–II*, edited by Michiel Streijger and Paul J. J. M. Bakker. Leiden: Brill.
Coronel, Luis. 1530. *Perscrutationes physice*. Lyons: Simon Vincent.
Diego of Astudillo. 1532. Quaestiones super octo libros Physicorum. Valladolid: Nicolás Tierri.
Henry of Ghent. 1983. In *Quodlibet IX*, edited by Raymond Macken. Leuven: Leuven University Press.
John of Jandun. 1551. Quaestiones subtilissimae super octo libros Aristotelis de Physico auditu. Venice: Iuntas.
Scotus, John Duns. 1997. *Quaestiones super libros Metaphysicorum Aristotelis. Libri VI–IX*, edited by Robert Andrews et al. St Bonaventure, NY: The Franciscan Institute.
Vernia, Nicoletto. 1505. *Quaestio de gravibus et levibus*. Venice: Antonio Pirro.

## *Secondary sources*

Åkerlund, Erik, Henrik Lagerlund, and Sylvain Roudaut. 2025. Blasius of Parma on Motion in the Sublunary World. In *The Mechanization of Nature: Matter, Body, and Motion in Blasius of Parma's Physics*, edited by Henrik Lagerlund, Sylvain Roudaut, and Erik Åkerlund, 72–92. Leiden: Brill.

Bakker, Paul J. J. M. 1999. *La raison et le miracle: Les doctrines eucharistiques (c. 1250–c. 1400): Contribution à l'étude des rapports entre philosophie et théologie. 2 vols. PhD diss.* Nijmegen: Radboud University.

Biard, Joël. 2022. *Hypothèse matérialiste et pensée radicale. La philosophie de la nature de Blaise de Parme.* Paris: Vrin.

Biard, Joël, and Aurélien Robert, eds. 2019. La philosophie de Blaise de Parme. Physique, psychologie, éthique. Florence: SISMEL–Edizioni del Galluzzo.

Cross, Richard. 2008. Accidents, Substantial Forms, and Causal Powers in the Late Thirteenth Century: Some Reflections on the Axiom '*Actiones sunt suppositorum*'. In *Compléments de substance: Études sur les propriétés accidentelles offertes à Alain de Libera*, edited by Christophe Erismann and Alexandrine Schniewind, 133–146. Paris: Vrin.

De Libera, Alain. 2006. Les actions appartiennent aux sujets: petite archéologie d'un principe leibnizien. In *«Ad Ingenii Acuitionem»: Studies in Honour of Alfonso Maierù*, edited by Stefano Caroti et al., 199–220. Louvain-la-Neuve: F.I.D.E.M.

Effler, Roy R. 1962. *John Duns Scotus and the Principle "Omne quod movetur ab alio movetur".* St Bonaventure, NY: The Franciscan Institute.

Ehret, Charles. 2021. *Agir en vertu d'un autre. Thomas d'Aquin et l'ontologie de l'instrument.* Paris: Vrin.

Federici-Vescovini, Graziella. 1991. Note sur la circulation en Italie du commentaire d'Albert de Saxe au *De caelo*. In *Itinéraires d'Albert de Saxe. Paris–Vienne au XIV^e siècle*, edited by Joël Biard, 235–251. Paris: Vrin.

Hassing, Richard F., and Edward M. Macierowski. 1992. Latin Averroes on the Divisibility and Self-Motion of the Elements. *Archiv für Geschichte der Philosophie* 74:127–157.

King, Peter. 1994. Duns Scotus on the Reality of Self-Change. In *Self-Motion: From Aristotle to Newton*, edited by Mary Louise Gill and James G. Lennox, 227–290. Princeton: Princeton University Press.

Lagerlund, Henrik, Sylvain Roudaut, and Erik Åkerlund, eds. 2025. *The Mechanization of Nature: Matter, Body, and Motion in Blasius of Parma's Physics.* Leiden: Brill.

Lang, Helen S. 1992. *Aristotle's Physics and Its Medieval Varieties.* Albany: State University of New York Press.

Maier, Anneliese. 1952. Die Ursache der Fallbewegung. In *Studien zur Naturphilosophie der Spätscholastik. Vol. 3: An der Grenze von Scholastik und Naturwissenschaft*, 143–182. Rome: Edizioni di Storia e Letteratura.

Pasnau, Robert. 2011. *Metaphysical Themes 1274–1671.* Oxford: Oxford University Press.

Pines, Shlomo. 1961. *Omne Quod Movetur Necesse est ab Aliquo Moveri*: A Refutation of Galen by Alexander of Aphrodisias and the Theory of Motion. *Isis* 52 (1): 21–54.

Roudaut, Sylvain. 2023. Can Accidents Alone Produce Substantial Forms? Twists and Turns of a Late Medieval Debate. *American Catholic Philosophical Quarterly* 97 (4): 529–553.

Roudaut, Sylvain. 2025. Luis Coronel's Theory of Impetus: The First 'Modern' Explanation of Free Fall Motion? *Early Science and Medicine* 30 (6): 609–639.

Sarnowsky, Jürgen. 1989. *Die aristotelisch-scholastische Theorie der Bewegung: Studien Zum Kommentar Alberts von Sachsen zur Physik des Aristoteles.* Münster: Aschendorff.

Solère, Jean-Luc. 2019. Scotus Versus Aquinas on Instrumental Causality. *Oxford Studies in Medieval Philosophy* 7:147–185.

Sylla, Edith D. 1993. Aristotelian Commentaries and Scientific Change: The Parisian Nominalists on the Cause of the Natural Motion of Inanimate Bodies. *Vivarium* 31 (1): 37–83.

Thijssen, Johannes M. M. H. 2009. The Debate over the Nature of Motion: John Buridan, Nicole Oresme and Albert of Saxony, with an Edition of John Buridan's *Quaestiones super libros Physicorum, secundum ultimam lecturam*, Book III, q. 7. In *Evidence and Interpretation in Studies on Early Science and Medicine: Essays in Honor of John E. Murdoch*, edited by Edith D. Sylla and William R. Newman, 186–210. Leiden: Brill.

Vucu, Simona R. 2018. *Henry of Ghent and John Duns Scotus on Self-Agency and Self-Motion: An Inquiry into the Medieval Metaphysics of Causal Powers. PhD diss.* Toronto: University of Toronto.

Wallace, William A. 1972. The Cosmological Argument: A Reappraisal. *Proceedings of the American Catholic Philosophical Association* 46:43–57.

Weisheipl, James A. 1985a. The Principle *Omne quod movetur ab alio movetur*. In *Nature and Motion in the Middle Ages*, edited by William E. Carroll, 75–98. Washington, D.C.: Catholic University of America Press.

Weisheipl, James A. 1985b. The Specter of Motor Coniuctus in Medieval Physics. In *Nature and Motion in the Middle Ages*, edited by William E. Carroll, 99–120. Washington, D.C.: Catholic University of America Press.

Zambiasi, Roberto. Forthcoming. From Metaphysical Postulates to Homeostatic Devices: An Early Fifteenth-Century Take on Substantial Forms as Efficient Causes. In *Arts Quodlibeta at the University of Vienna: The 1429 Quodlibet under Urban of Melk*, edited by Chris Schabel and Ota Pavliček. Turnhout: Brepols.

**Sylvain Roudaut** (PhD, 2017) is Senior Researcher at the CNRS (SPHERE, Paris). He previously held research and teaching positions at Stockholm University and KU Leuven. His research focuses on the interaction between philosophy, science, and mathematics in the Middle Ages. He published a monograph on quantification techniques in the Middle Ages (*La mesure de l'être*, Brill, 2021) and co-edited several volumes on medieval natural philosophy and metaphysics, including recently *The Ontology of Artifacts* in the Middle Ages (*Philosophies*, 2022) and *Hylomorphism into Pieces* (Palgrave Macmillan, 2024).

# Chapter 4
# *Higgs Boson Blues*. Revisiting Nicholas of Autrecourt's Atomism

**Gustavo Fernandez Walker**

**Abstract** Nicholas of Autrecourt's defence of atomism has been the object of several studies by many scholars. My aim in this paper is twofold: first, in response to the invitation extended by this conference, I would like to go through the Aristotelian foundations of Autrecourt's interpretation of atomism—a methodology that can be characterised as "climbing on the shoulders of giants, and subsequently removing the ladder." Second, I will focus on the notion of *situalitas* in order to trace the problems it is called to solve, namely how objects with mass are constituted from massless atoms, and the new issues that arise from the coinage of this new term in fourteenth-century discussions about natural philosophy.

**Keywords** Nicholas of Autrecourt · Medieval atomism · *Situalitas*

## 4.1 Introduction

Nicholas of Autrecourt's defence of atomism has been the object of several studies.[1] In most, if not all, cases, one concept is identified as crucial to understanding his position: *situalitas*. In fact, this appears to be a neologism coined by Autrecourt: there are instances of the term as an adverb (*situaliter*) or adjective (*situalis*) among some of his contemporaries,[2] but the further step of transforming the concept into a noun, with the typical suffix (*-itas*) corresponding to abstract terms, appears to have been taken by Autrecourt himself.

[1] For an overview of the studies devoted to this issue, see, for instance, Grellard and Robert (2009), Kaluza (1998), reprinted in Kaluza (2023).

[2] See for instance Robert 2009, Wood 2009

G. F. Walker (✉)
FLoV – Göteborgs Universitet, Göteborg, Sweden
e-mail: gustavo.fernandez.walker@gu.se

 
E. Jung et al. (eds.), *The Forgotten Giants: Overlooked Revolutions in Natural Philosophy*, Studies in History and Philosophy of Science 70,
https://doi.org/10.1007/978-3-032-34008-5_4

Already in 1974, John E. Murdoch pointed out[3] that many fourteenth-century atomists relied on *ad hoc* solutions to bypass one of the main objections levelled against atomists since Aristotle, namely how it is possible for dimensionless atoms to make up dimensional objects. There is no problem, in principle, with *ad hoc* solutions, if they fulfil their task and offer, indeed, a solution. My aim in the present chapter is to assess whether the notion of *situalitas* rises up to the challenge and, if so, how this task is accomplished.

As a side note regarding the picture of "forgotten giants" proposed as a topic for this conference, the case of fourteenth-century atomists seems to be slightly different: their main source for their vindication of atomism was Aristotle, and their goal was to achieve a defence of atomism based on a refutation of his refutation. In a way, they were climbing on the shoulders of a giant in the hope of removing the ladder once the task was accomplished.

The title of this paper is taken from a 2013 song by Nick Cave. In it, a man drives to Geneva remembering some old lost love while he passes near the Large Hadron Collider. The decades of discussion about the existence of the Higgs boson, an elementary particle that would explain how particles acquire mass are, in a way, an echo of fourteenth century discussions about the mechanism by which massless atoms constitute objects with mass.

## 4.2 First Problem: Contact Between Indivisibles

In the chapter of his *Exigit ordo* devoted to indivisibles, Autrecourt expresses that atoms, themselves dimensionless, can however get together and form dimensional objects because they possess what he calls their own *situalitas*.[4] In his extant writings, he never offers a detailed account of what this *situalitas* actually is. It is, however, clear what role it plays in his own explanation of how atoms behave: it is the concept brought up to respond to some of the main objections levelled against atomism.

The first objection is the famous argument that Aristotle brings up in *Physics* IV. Autrecourt summarises it in the following manner:

> Aristotle has another argument, the gist of which is as follows: If a continuum were composed of points, either the points would touch one another or not. If not, a continuum or a *contiguum* would never be made from them. If they do, then either a whole point touches a whole point, or part touches part. But part does not touch part, since then points would be divisible, which was denied originally. And, if a whole touches a whole, it does not make it any larger, which is against the hypothesis.[5]

[3] Murdoch (1974, 17): "Il est évident que l'invention de ces nouvelles méthodes de contact entre indivisibles était, á proprement parler, *ad hoc*."

[4] As Grellard points out, Autrecourt's atomism "is primarily physical and only secondarily mathematical" (2009, 108). His understanding of the atom is conflated with the notions of *indivisibilia* and Aristotle's *minima naturalia.*

[5] Nicolaus de Ultricuria (2024, 160): "Alia quidem ratio Aristotelis, cuius virtus sequitur: Si continuum componeretur ex punctis, aut puncta se tangerent, aut non. Si non, tunc ex eis numquam

Since at least Aristotle, explaining how indivisible elements stay in contact with each other has been the Achilles' heel of atomists. How is an extension created by this union of non-dimensional atoms? "A whole touching a whole" seems to imply that two atoms occupy the same space, something which is either impossible or, even if one were to accept it, would not render the whole larger, given that the space occupied would remain the same. For Nicholas of Autrecourt, introducing the concept of *situalitas* is the means to explain the way in which a dimensionless whole can be in contact with another dimensionless whole, bypassing these unwanted consequences:

> But one could reply thus to this argument (*ad hoc*): 'A whole touching a whole' can only be understood in one way, namely that nothing of one is related to the other mediately, but the whole of one is related immediately to the whole other. And this kind of contact *is not possible in bodies* because one body is always related to another mediately regarding their remote parts, so that bodies do not touch one another except by means of the parts that are closer together. In this way, and *contrary to what happens with bodies*, it is possible to say that a whole point touches another point as a whole, in the manner explained. Someone could <also> understand (although perhaps this understanding does not arise from the literal meaning of the words) that <'a whole touching a whole' means that> one whole is not outside of the other. I reject this interpretation, for in this second way a whole touching a whole would not make it larger; but in the first way it would make it larger. From which it can be said that a point has its own *situalitas* and its own mode of being. Therefore, if points are put together and retain their own *situalitas*, it is senseless to say that they will not constitute something larger.[6]

This passage contains a significant amount of information to analyse, as it introduces the concept of situalitas in the *Exigit ordo* for the first time (the only other instance appears shortly thereafter, addressing a separate argument, which we will examine presently). Autrecourt seems determined to remind the reader that one cannot apply the same explanations to bodies and atoms: twice in the passage he explicitly states the difference between what is to be understood by "contact" in either case. The kind of contact referred to in this discussion ("a whole touching a whole") creates an apparent refutation of atomism, precisely because it relies on the (false) assumption that points/atoms behave like composite bodies. The Aristotelian objection in *Physics*

---

fieret continuum neque contiguum. Si sic, aut ergo totum punctum tangit totum vel pars partem. Non pars partem, quia tunc essent divisibilia, quod in primis negabatur. Si totum tangit totum, tunc non reddit ipsum maius, quod est contra positum." I use here Kennedy's translation (Nicholas of Autrecourt 1971, 73), slightly modified when necessary. Modern translations available of the treatise are in English (Nicholas of Autrecourt 1971, trans. Kennedy et al.), Italian (Nicola di Autrecourt 2009, trans. Musu), and German (Nicolaus de Ultricuria 2024, trans. Berger).

[6] Nicolaus de Ultricuria (2024, 160): "Ad hoc autem posset sic responderi: totum tangere totum uno <modo> potest intelligi, id est, quolibet sui, ita quod nihil ipsius mediate se habeat ad aliud, sed totum ipsum se habet immediate totum ad ipsum. Et iste tactus *non est possibilis in corporibus*, quia semper secundum partes remotas unum corpus se habet mediate ad aliud et se non tangunt nisi mediantibus partibus propinquis. Isto modo *ad differentiam corporum* potest dici quod punctus se toto alium tangit secundum modum positum. Alio modo posset quis intelligere, licet forsan intellectus iste non habeatur de virtute sermonis, quod totum, id est unum, non est extra aliud, et istum intellectum nego. Lsto enim secundo modo totum tangens totum ipsum non faceret majus, sed primo modo reddet ipsum majus. Unde punctus habet propriam *situalitatem* et proprium modum essendi, et ideo si approximarentur puncta remanentibus *situalitatibus* propriis, tunc est inintelligibile quod non facient majus."

VI.1 (231b15–232a20) simply states that indivisibles either are, in fact, divisible (if they are in contact with one another part against part), or otherwise incapable of making up a larger compound (if in contact whole to whole).

Autrecourt's solution to the Aristotelian dilemma hinges on this equivocation. In contrast to physical bodies, dimensionless atoms can indeed have a complete contact with one another while simultaneously not occupying the same place. For Autrecourt, this phenomenon is feasible because an indivisible entity, which cannot occupy any place—this characteristic pertains to composite bodies rather than atoms—still maintains a certain relationship with space: this is referred to as its *situalitas*.

Incidentally, this offers a possible answer to a question usually asked by readers of Autrecourt: Why did he feel compelled to create the neologism *situalitas* when the more common term *situs* was already available? One could argue that *situs* is a concept that applies to the composite body; whereas *situalitas*, with its suffix typical of abstract nouns, refers to the kind of relation that a dimensionless atom establishes with space. Thus, as the Aristotelian objection stated, multiple atoms do occupy the same *situs* (the actual space occupied by the composite made up by those atoms), each retaining however their own *situalitas*, which serves as the key to defuse the Aristotelian threat. Taking the lead offered by Z. Kaluza[7] about Autrecourt's affinity with Scotistic positions with regard to formal reasons, we can in a way understand each atom, with its own *situalitas* (the formal principle that allows a dimensionless atom to take its position in a composite), occupying the same place (*situs*) as the composite, in a way similar in which one and the same common nature, for Scotus, can be present in several places simultaneously, thanks to the *haecceitates* that are really identical but formally distinct with it.[8] Incidentally, another feature shared by these neologisms (the Scotist *haecceitas* and the Ultricurian *situalitas*) is their capacity to take scribes by surprise: the only extant manuscript of the *Exigit ordo* (Oxford, Bodleian Library, Ms. Canon. misc. 43, fol. 1ra–24va) reads, in the passage quoted above, *situliatatem* (!) instead of *situalitatem*, much like scribes copying texts by Scotus recurred to *hereitas*, *habilitas* or other terms that, to the eyes of the scribe, were more familiar than this abstraction.[9]

Christophe Grellard rightly points out Autrecourt's motivation for this complex theory.[10] As the ultimate elements of reality, atoms are not perceivable, which means that abstraction and imagination are required to arrive at an understanding of their mode of being and operations. The atom is, for Autrecourt, a metaphysical concept, neither mathematical nor strictly physical. Certainly, the philosophers' goal is to give an accurate account of natural phenomena. But the level of abstraction required to

[7] See Kaluza (1998), reprinted in 2025 (esp. 83 and 103).

[8] Autrecourt shows familiarity with this distinction in his discussion on the eternity of things, cf. Nicolaus de Ultricuria (2024, 146–148): "intellectus per diversitatem situs ponit differentiam, nam evidens est intellectui quod una res non potest esse in pluribus locis simul. [...] Verum est quod tale non potest esse in pluribus locis, sed quin ratione speciei possit, non est verum."

[9] See Andrews (2010).

[10] See Grellard (2004).

achieve that goal forces them to come up with these arguments and, if necessary, to develop new concepts that can explain them.[11]

## 4.3 Second Problem: The *Continuum*

After refuting this first argument against atomism, Autrecourt offers another possible argument against it:

> Another objection is presented by certain others: If a continuum were composed of points, the quantified would be made of the non-quantified, since a continuum is quantified and a point is not quantified. Now, either you understand by the quantified *(a)* the composite itself or its extension—and then a point is not quantified, although according to this interpretation the quantified can be made from the non-quantified just as the composite from what are incomposite <elements>—; or you understand by the quantified *(b)* that which has its own *situalitas* and a circumscribed being—and thus I grant that a point can be said to be quantified, and you will not then be able to conclude from this that the quantified is made from the non-quantified.[12]

Autrecourt considers two possible ways out of the problem of making up a *quantum* out of *non-quanta*. Both involve dealing with the notion of *quantum*: if by *quantum* you understand anything that has an extension, then the composite is a *quantum* and the atom is not. If that were the case, the only alternative would be to consider that, analogous to a composite substance being made up of simple elements, you would have to consider a quantified extension being made up of extensionless atoms. Since there seems to be no justification to accept this analogy, an alternative presents itself: granting some degree of quantity to dimensionless atoms.

For this argument to work, Autrecourt summons again his newly discovered concept of *situalitas*. In this context, *situalitas* is offered as a sufficient condition to identify a *quantum*. The argument stops there: if *situalitas* had been posited as a necessary condition as well as a sufficient one, then bodies would have to possess their own *situalitas*. And this, Autrecourt does not seem to accept: *situalitas* is the condition required by atoms to become part of a dimensional body. If anything, the body has a *situs*. And then if one accepts the possession of *situalitas* a sufficient condition to acknowledge a *quantum*, then atoms are *quanta*, and the objection loses all its power: a composite body made up of atoms wouldn't be a case of a *quantum* being made up of *non quanta*; in fact a *quantum* (the composite) would then be made up of *quanta* (the atoms).

---

[11] See Grellard (2004, 190–192).

[12] Nicolaus de Ultricuria (2024, 162–164): "Et quibusdam aliis <omissis> fit ratio quaedam: Ex non quanto fieret quantum, si ex punctis continuum componeretur, quia continuum quantum est, punctum vero non est quantum. Aut ergo per quantum intelligis ipsum compositum seu ipsam extensionem, et sic punctum non est quantum. Tamen secundum istum intellectum ex non quanto potest fieri quantum, quemadmodum ex non compositis compositum. Vel per quantum intelligis habens propriam situalitatem et esse circumscriptive, et sic concedo quod punctum potest dici quantum, et tunc non poteris secundum hoc concludere ex non quanto fieri quantum." I follow here Kennedy's translation (Nicholas of Autrecourt 1971, 74), with some modifications.

Naturally, in this case, the reasons by which the composite and its atoms can be considered *quanta* are different: for the composite, there doesn't seem to be an inconvenience. In the case of atoms, however, it was first required of us that we grant some degree of quantity to them. And this is the work that the notion of *situalitas* is called up to do: it is precisely because it is endowed with this *situalitas* that an otherwise dimensionless atom can be considered a certain kind of *quantum*, as the previous discussion suggested. Thus considered, an atom can, in conjunction with other atoms, each with their own *situalitas*, conform a quantified composite.

Thus stated, this sounds precisely as the type of *ad-hoc* solution that Murdoch cautioned about.[13] But Autrecourt seems to be doing something more than just coining an *ad-hoc* neologism simply to defuse an objection. It is true that there is no definition of the concept in the *Exigit ordo*. But from its use in the treatise, in the arguments just reviewed, it seems possible to draw some conclusions.

## 4.4 Conclusion

At first glance, one can notice that, in both arguments in which the concept of *situalitas* is introduced—one dealing with contact between indivisibles, the other with quantity and the continuum—the notion is introduced with the addition of a second concept: *modus essendi* in the first argument, *esse circumscriptive* in the second. In either context, when the notion is introduced, it is always paired with one of these: *situalitas* et *modus essendi*, *situalitas* et *esse circumscriptive*. Are these expressions equivalent? In other words, do these expressions refer to three, two, or one concept(s)?

Given that *situalitas* is present in both arguments, once paired with *modus essendi*, once with *esse circumscriptive*, at first there seems to be three different concepts involved. However, the identity of one of the conjuncts in either expression (*situalitas*) could induce us to consider the other two as equivalent. In that case, Autrecourt would be discussing here two concepts: *situalitas* on the one hand, *modus essendi*/*esse circumscriptive* on the other. Finally, if the particle *et* is considered here in its epexegetic function, we could consider all these as three different labels for one single concept: *situalitas*, i.e., *modus essendi*, i.e., *esse circumscriptive*.[14]

This last possibility is the easiest to discard: the epexegetic use of the particle *et* is rare in scholarly texts like the *Exigit ordo*, and more apt for a poetic or rhetorical context. The first possibility is also difficult to defend: it implies that Autrecourt,

[13] In fact, what seems to emerge from Autrecourt's positions on this issue, is the understanding of an atom as a seemingly paradoxical dimensionless *quantum*.

[14] Available modern translations seem to fall between the first and second possibility: "its own position and a circumscribed being" (Nicholas of Autrecourt 1971, trans. Kennedy et al.); "eine eigene Lage und ein Sein durch Begrenzung" (Nicolaus de Ultricuria 2024, trans. Berger); "una propria posizione spaziale ed un proprio modo di essere a prescindere di tutto il resto" (Nicola di Autrecourt 2009, trans. Musu). Notice how, in this last case, *esse circumscriptive* seems to be understood more as a *modus intelligendi* than as a *modus essendi*, i.e., *circumscriptive* is taken as modifying *intelligis* rather than *esse*.

not content with introducing the notion of *situalitas* without any further explanation, proceeds later to pair it with two separate concepts leaving the reader with the extra task of explaining the relation of *situalitas* with either of them. Thus, option number two seems more plausible: *esse circumscriptive* in the second argument seems to elaborate on the pairing presented in the first, namely that of *situalitas* + *modus essendi*. How, then, to understand this mode of being?

In the first argument, *situalitas* and *modus essendi* are not only presented in a conjunction: they are both introduced with the adjective *proprius*. Atoms, then, possess their own *situalitas* and their own mode of being. But in the second argument, the adjective *propria* is reserved only to *situalitas*, not to *esse circumscriptive*. This shouldn't be a surprise: circumscribed being seems precisely to limit the atom so that it is distinguished from any other. In a way, *proprium modum essendi* and *esse circumscriptive* are, if not necessarily equivalent, indicative of the singularity of each atom. Once again, and in a manner not incompatible with a Scotistic framework, the *situalitas* of an atom is the condition of possibility of it becoming a part of a *quantum* that is situated (i.e., that has a *situs*).[15]

Furthermore, medieval discussions regarding the location of angels offer some insight into the possible use of *esse circumscriptive* in Autrecourt. Like atoms, angels are non-dimensional, and so their relation to space becomes a disputed issue among theologians. The notion of *esse circumscriptive* when discussing the location of angels was the way to block, at the same time, the possibility of angels being nowhere and/or everywhere. Even if their place was not determined, they had to be in a determinate place.[16]

In the context of discussion on the continuum, Autrecourt seems to summon this notion of *esse circumscriptive* to carry out a similar task. Atoms, being dimensionless, cannot occupy a space. But they have a mode of being that makes them capable of being a part of a *quantum* that is itself located. This *situalitas* is what sets atoms apart from each other: in that sense it can be considered their *propria situalitas*, and thus their mode of being makes them potentially capable of occupying a space when joined to other, equally dimensionless atoms.

But then the notion of *situalitas* reveals itself as much more than a mere *ad hoc* resource to cancel a difficult objection. It reveals itself as a cornerstone of Autrecourt's atomistic program. One of the condemned propositions from the *Exigit ordo* states that *In rebus nature non est nisi motus localis*. How are we to understand local motion of atoms if atoms themselves are not in any way located? The *situalitas* proper to each atom is then the condition of possibility of it being locally situated within a composite. Such is their circumscribed mode of being: not fixed in a place, but capable of occupying the space of the continuum of which they are a part of.

As a final remark, one should observe that this reading of Autrecourt is, as he himself describes his program, merely probable. Moreover, one should not forget the status of *brouillon* (Z. Kaluza *dixit*)[17] of the *Exigit ordo*, most evident in those

[15] See for instance Suárez-Nani (2008).

[16] See the articles collected in Iribarren and Lenz (2008).

[17] Kaluza (1995, 170).

passages in which the author seems to have left "notes to self." H. Berger points out in his edition the various places in which Autrecourt closes a discussion with the following *caveat*: *Et hic considera amplius.*[18] It serves as well as a friendly reminder to us, his readers.

**Acknowledgements** Research for this paper was carried out within the project *Reassessing Aristotelian Science*, funded by the Knut and Alice Wallenberg Foundation and led by Ana María Mora Márquez (Lunds Universitet). I am grateful for all the comments received both during the conference at Lodz as well as from readers of earlier drafts of this paper.

## References

Andrews, Robert R. 2010. Haecceity in the Metaphysics of John Duns Scotus. *Archa Verbi. Subsidia* 5: 151–161.

Grellard, Christoph. 2004. Les présupposés méthodologiques de l'atomisme: la théorie du continu chez Nicolas d'Autrécourt et Nicolas Bonet. In *Méthodes et statut des sciences à la fin du moyen âge*, edited by Christoph Grellard, 181–199. Lille: Presses Universitaires du Septentrion.

Grellard, Christoph, and Aurélien Robert, eds. 2009. *Atomism in Late Medieval Philosophy and Theology*. Leiden: Brill.

Iribarren, Isabel, and Martin Lenz, eds. 2008. *Angels in Medieval Philosophical Inquiry: Their Function and Significance*. Aldershot: Ashgate.

Kaluza, Zénon. 1995. *Nicolas d'Autrécourt, ami de la vérité*. Paris: Klincksieck.

Kaluza, Zénon. 1998. Les catégories dans "l'Exigit ordo". Étude de l'ontologie formelle de Nicolas d'Autrécourt. *Studia Mediewistyczne* 33: 97–125.

Kaluza, Zénon. 2023. *L'ontologie de Nicolas d'Autrécourt*. Leiden: Brill.

Murdoch, John E. 1974. Naissance et développement de l'atomisme au bas Moyen Âge latin. In *La science de la nature: théories et pratiques*, edited by Guy H. Allard and Jacques Ménard, 11–32. Montréal: Bellarmin.

Nicholas of Autrecourt. 1971. *The Universal Treatise*. Translated by Leonard A. Kennedy et al. Milwaukee: Marquette University Press.

Nicola di Autrecourt. 2009. *Il «Trattato»*. Translated by Antonella Musu. Pisa: Edizioni ETS.

Nicolaus de Ultricuria. 2024. *Exigit ordo*. In Nikolaus von Autrécourt, *Allgemeine Abhandlung gegen die Aristoteliker. Lateinisch–Deutsch*, edited and translated by Harald Berger. Freiburg im Breisgau: Herder.

Robert, Aurélien. 2009. William Crathorn's Mereotopological Atomism. In *Atomism in Late Medieval Philosophy and Theology*, edited by Christoph Grellard and Aurélien Robert, 127–162. Leiden: Brill.

Suárez-Nani, Tiziana. 2008. Angels, Space and Place: The Location of Separate Substances According to John Duns Scotus. In *Angels in Medieval Philosophical Inquiry. Their Function and Significance*, edited by Isabel Iribarren and Martin Lenz, 89–111. Aldershot: Ashgate.

Wood, Rega. 2009. Indivisibles and Infinities. In *Atomism in Late Medieval Philosophy and Theology*, edited by Christoph Grellard and Aurélien Robert, 39–64. Leiden: Brill.

[18] Nicolaus de Ultricuria (2024, 322, n. 267).

**Gustavo Fernandez Walker** is Assistant researcher at the Department of Philosophy, Linguistics and Theory of Science (FLoV) at the University of Gothenburg, Sweden. He holds a Ph.D. in Philosophy (University of Buenos Aires, Argentina) and a Ph.D. in Philology and Hermeneutics of the philosophical text (Università del Salento, Italy). He was visiting researcher at the Research Center for Comparative History of Religious Orders (FOVOG, Technische Universität Dresden) and associate professor of Medieval Philosophy at the University of San Martín, Argentina. His research is mainly concerned with Nicholas of Autrecourt and the Medieval reception of Aristotle's Organon.

# Chapter 5
# Commitment Issues? Marsilius of Inghen, Nominalism and the Semantics of Necessarily Empty Terms in the Late-Fourteenth Century

**Graziana S. Ciola**

**Abstract** Marsilius of Inghen (ca. 1341–1396) was well known in his time and in the following centuries especially as a natural philosopher. His influence in the history of science is well known. But his impact on the history of logic and metaphysics is just as important, notably for the shaping and the historical development of broader philosophical categories such as the "modern way" (*via modernorum*) and "nominalism" as we understand it. This article examines the apparent tension between nominalist commitments and the acceptance of non-existent objects in late-medieval logic, focusing particularly on Marsilius of Inghen's semantics and its enormous but understated impact on the history of logic. I argue against the common assumption that nominalism's ontological parsimony necessarily excludes reference to *impossibilia* and should even exclude any reference to *possibilia.* Instead, I claim that Marsilius of Inghen and his followers developed sophisticated semantic frameworks that could accommodate empty terms while maintaining their anti-realist commitments in meaningful ways. Through an analysis of supposition theory, ampliation, and the treatment of *imaginabilia*, I show how Marsilius of Inghen and his followers created a coherent system that allowed meaningful discourse about non-existent and even impossible objects. This examination highlights fundamental differences between contemporary and medieval conceptions of nominalism and suggests that the ontological commitments of medieval nominalists were more nuanced than typically assumed. This study traces the development of these theories through the works of John Buridan, Albert of Saxony, and Marsilius of Inghen, examining their reception in later commentaries and their implications for our understanding of medieval semantic theory.

**Keywords** Medieval logic · Nominalism · Empty terms · Supposition theory · Marsilius of Inghen · John Buridan · *Impossibilia* · Ampliation

G. S. Ciola (✉)
Radboud University, Nijmegen, The Netherlands
e-mail: graziana.ciola@ru.nl



E. Jung et al. (eds.), *The Forgotten Giants: Overlooked Revolutions in Natural Philosophy*, Studies in History and Philosophy of Science 70,
https://doi.org/10.1007/978-3-032-34008-5_5

## 5.1 Introduction[1]

Both during his lifetime and the following centuries, Marsilius of Inghen (ca. 1341–1396) was especially renowned for his contributions in natural philosophy. His *Quaestiones in libros De generatione et corruptione* remained extremely popular and influential throughout the fifteenth century and sixteenth.[2] For example, even though Marsilius is just one of the fourteenth-century authors young Galileo Galilei still read and quoted more or less in passing through his *Iuvenilia*, the commentary on *De generatione et corruptione* is a text with which he actively engaged in order to polemicise against the doctrines advanced there.[3] But Marsilius' role in the history of logic and metaphysics may be just as impactful as his reception in the history of natural philosophy and scientific thought. Here, I argue that Marsilius is a pivotal figure for shaping the historical development of broader philosophical categories—such as the "modern way" (*via modernorum*) and the notion of "nominalism" associated with it.

"Nominalism" is a contested label in the historiography of medieval philosophy, especially when applied to fourteenth-century authors. These thinkers did not describe themselves as "nominalists" or even as "moderns"; rather, the label was retrospectively adopted by their fifteenth-century followers.[4] The situation is further complicated by the fact that, in the early twentieth century, logical empiricists also appropriated the term "nominalism" to characterise their own ontologically parsimonious and anti-metaphysical positions—occasionally even mentioning previous historical iterations of nominalism.[5]

Nominalism generally demands ontological parsimony: a nominalist avoids commitment to universals, complex facts,[6] and unrealised possibilities.[7] Fourteenth-century nominalism, however, was more nuanced. In particular, it was never a wholesale rejection of abstract entities, as contemporary nominalism often is.[8]

[1] Earlier versions of this paper were presented at the St Andrews *Metaphysics and Logic Seminar* and at the *American Philosophical Association Central Division annual meeting.* I am especially grateful to Franz Berto, Milo Crimi, Bob Pasnau, Stephen Read, and Greg Restall for their comments and insights.

[2] Cf. Hoenen (2025), Walz and Düchting (2008), Thijssen (1999).

[3] For example, in the *Tractatio de caelo* (q. 5), Galileo mentions in passing Marsilius' *Quaestiones super Metaphysicam*, but he engages in detail with Marsilius' *Questiones super De generatione et corruptione* in his *Tractatus de elementis* (q. 3)—see Galilei (1929, respectively 72 and 167).

[4] Cf. e.g., Courtenay (1987). Despite being dated, the collection in Zimmermann (1974) still offers a rich and broad overview on these matters between the twelfth century and the fifteenth.

[5] See, for example, Carnap (1931).

[6] Especially between the thirteenth and the fourteenth century, nominalist views normally express forms of terminism in logic in explicit contrast with modism, which is metaphysically robust and—claiming that there are ways of being, matching ways of understanding and ways of signifying—expresses fundamentally realist positions. See Courtenay (1987), Dutilh Novaes (2011, esp. 1232–1234: "The Age of Terminism."). Theories of *complexe significabilia* seem uncomfortably close to the modist framework, even—or especially—when endorsed by someone such as Gregory of Rimini, who stands otherwise firmly among the *moderns* and remains fundamentally noncommittal. See, e.g., Schabel and Girard (2022).

[7] See e.g., Pasnau (2025).

[8] Cf. Rodriguez-Pereyra (2019), Cowling and Giberman (2025).

Can one be a nominalist while endorsing discourse about non-existent objects? While this may seem paradoxical from a contemporary perspective, it is essential for understanding late-medieval semantic theories. Marsilius of Inghen's semantics can provide a meaningful treatment of necessarily empty terms, including references to impossible objects. Although this may appear un-nominalist, I argue that Marsilius was both a nominalist and a Meinongian, and reasonably so.

In the present contribution, I shall proceed as follows.

First, I outline three key divergences between medieval and contemporary nominalism, arguing against projecting contemporary constraints—especially regarding modal semantics and abstract entities like concepts—on the medieval framework. This legitimises applying "nominalism" to Marsilius of Inghen's semantics while suggesting that medieval nominalism's non-conformity to contemporary paradigms offers valuable insights for current semantic and metaphysical debates.

Second, I examine fourteenth-century nominalists' shared semantic framework and its challenges with empty terms. Analysing Buridan's and Marsilius' alternative strategies, I show how each preserves nominalist commitments differently—Buridan grounds concepts in empirically efficacious particulars; Marsilius adopts a more theory-internal approach.

Finally, I trace the late-fourteenth and fifteenth-century reception of Marsilius' theory through John Dorp' *Summulae* and the Hagenau Commentary, demonstrating how later nominalists integrated causal and epistemological concerns into their semantic analysis. While Marsilius separated formal semantics from concept formation, later commentators united these dimensions, explaining how the intellect understands *impossibilia* through compositional operations on concepts from possible objects. This integration elaborated Marsilius' framework through the epistemological engagement that Buridan had favoured, shaping subsequent nominalist semantic theory. Throughout, I argue that the tension between ontological parsimony and making sense of necessarily empty terms expresses the sophistication of medieval nominalist semantics rather than its inconsistency.

## 5.2 Contemporary vs. Medieval Nominalism: A (Digest) Tale of Two Anti-Realisms

In their broadest characterisations, all nominalisms are forms of anti-realism—i.e., what makes a philosophical stance essentially "nominalist" is rejecting the *reality* of something. What things are rejected and the strength of that rejection may vary. In this broad sense, both medieval and contemporary nominalism are anti-realisms of some kind. But, despite the many conceptual overlaps and a loose relation of historical derivation between them, medieval and contemporary nominalism radically and most evidently diverge on at least one fundamental point, i.e., their assessment of abstract objects.

In metaphysics and in semantics, the majority of the medieval philosophical stances that we commonly qualify as "nominalism" are more precisely forms of "conceptualism." They could be distinguished with various degrees of nuance, but overall, they all tend to accept concepts and abstract entities at least in some capacity.[9]

Instead, contemporary nominalism is typically defined by its rejection of abstract objects, with the requirement that everything is particular and concrete rather than abstract.[10] This translates into an exclusive commitment to spatiotemporally existing entities and the rejection of anything that would be causally inert. In the early iterations of contemporary nominalism, the main philosophical concern seemed to stem from a strongly empiricist stance and a general diffidence towards metaphysics—or at least some types of metaphysics.[11] This was the case with the logical positivists in the last century, as well as with those philosophers who positioned themselves in their wake and extended a strong influence on the so-called analytical tradition. Among others, W.V. Quine is the obvious name to mention by way of example due to his dual dislike of quantified modal logic as entailing essentialism and of intrinsically contradictory objects—such as the round square cupola on Berkeley College—as fundamentally inconceivable.[12] These have been common features of contemporary forms of nominalism and nominalistic approaches, in addition to a marked emphasis on causality as the discriminant for what should count as a concrete, non-inert object.[13] This focus reflects concerns about causal isolation and epistemological access to abstract objects.

In turn, these concerns reflect metaphysical assumptions medieval philosophers did not share. Medieval thinkers operated within an Aristotelian framework where causation, space, and time were understood differently, and immaterial entities like God and angels were causally effective.[14]

Medieval nominalism is best known as one of the contestants in the Debate(s) on Universals. In this sense, nominalism was a metaphysical stance primarily interested in rejecting the reality of archetypes and common properties inside individual things rather than abstract objects generally.[15] The rejection of Universals is one of those

---

[9] See, e.g., Klima (2022), Spade, Panaccio, and Pelletier (2025).

[10] For a broad introduction outlining the nuanced differences between the rejection of abstract objects in general and the rejection of general abstract objects (i.e., universals), see Guigon (2019).

[11] For a broad overview of nominalist-leaning positions in the contemporary analytical tradition shaped by the logical empiricist movement, see Creath (2023).

[12] Cf. respectively Quine (1948, 1953).

[13] This aspect is clearly and carefully detailed in Rodriguez-Pereyra (2019).

[14] For an overview on causation, see White 2018). As for space, time and motion, historians of science traditionally attribute the formulation of these notions as absolute to Newton, even if ancient and medieval precursors are discussed by the scholarship. For an overview, see among others: Disalle (2006). The literature on various aspects concerning efficient immaterial beings across various branches of medieval philosophy is extensive. A stimulating overview of material and immaterial items in the long Middle Ages can be found in Pasnau (2011).

[15] The "Problem of Universals" is probably *the* most discussed philosophical discussion in medieval philosophy. To my knowledge, the most useful and regularly updated overview of the matter is to this date: Klima (2022).

baseline features of "nominalism" that is shared by medieval and contemporary iterations, but certainly it is not the full extent of what medieval nominalism was and how it was expressed in several areas of philosophy. Crucially, medieval nominalists did not thereby commit themselves to rejecting all abstract entities. This allowed them to maintain robust theories about mental content, mathematical objects, and even impossible entities without compromising their anti-realist stance.[16]

Projecting contemporary nominalism onto medieval philosophers would be a category mistake given their different understanding of space, time, and causation. Especially causal relations were understood as relatives (*ad aliquid*) that could thus be "shaved away" by the nominalist—and especially Ockhamist—reductionist reframing of the categories.[17] Consequently, being non-causally inert does not seem a suitable requirement either. This different metaphysical background allowed medieval nominalists to maintain their anti-realist commitments while developing semantic theories that modern nominalists might find ontologically objectionable. The question is not whether medieval nominalists were consistent by contemporary standards, but whether they were reasonable—given their own constraints, theoretical commitments and historical context.

Fourteenth-century nominalism was fundamentally conceptualist rather than eliminativist. Ockham, Buridan, Albert of Saxony, and Marsilius accepted concepts, mental terms, and complex thoughts as mental products without metaphysical independence. This conceptualist approach allowed strong theories of meaning without any commitment beyond individual items. Conceptualism and the fourteenth-century emphasis on the primacy of mental language (mentalism)[18] allow to think of mathematical and scientific items as mental constructions rather than independently existing abstract objects. Moreover, conceptualism, in the peculiar declension it assumed in mid-fourteenth century nominalism, created room for sophisticated treatments of problematic cases like empty terms and impossible objects in terms of complex concepts and mental operations.[19] This conceptualist framework also explains why at least some fourteenth-century nominalists did not see their semantic theories as ontologically problematic in the way contemporary nominalists would. Overall, mental entities were not abstract objects in the contemporary sense but rather concrete psychological occurrences—particular mental acts or states that existed in particular minds at particular times, namely while instantiated.[20]

---

[16] I previously addressed Marsilius of Inghen's deployment of *imaginabilia* as mere tools in semantics and in the logical analysis of semantic statements in Ciola (2018a, 2019, 2020).

[17] See, e.g., Klima (1999).

[18] Buridanus (2005, 1.1.6, 16–17).

[19] Notoriously, Buridan analyses away terms apparently signifying items that do not exist and cannot possibly exist as complex concepts ultimately made up of perfectly possible simple concepts. See Sect. 6 below.

[20] The requirement of current instantiation for any term or conglomeration of terms—be they mental, spoken or written—is one of the staples of the semantic presupposition we label as tokenism, which is a widespread and strict condition in late-medieval logic. On tokenism, see e.g. Klima (2004), Dutilh Novaes (2005), Ciola (2026).

Against this background, accepting complex mental structures involving impossible combinations of concepts is no more ontologically problematic than accepting complex mental acts. However, epistemic, ontological and semantical complications still emerge, and they do not do so only due to seemingly stronger treatments—such as Marsilius'—of necessarily empty terms as having signification in a fairly unitary sense.[21] Instead, some more fundamental issues arise from the basic semantic framework that these authors share. This framework is what we take as the common ground to define the borders of logical and semantical nominalism in the fourteenth century.

## 5.3 The Problem of Empty Terms in Nominalist Semantics

The semantic framework shared by fourteenth-century nominalists—what Terry Parsons called BOB, the "Best of Buridan"—represents a common ground between Buridan, Albert of Saxony, and Marsilius.[22] This framework, rooted in the Aristotelian semantic tradition, makes empty terms especially problematic.[23]

BOB endorses mentalism and tokenism. Concepts are mental terms, which directly and naturally signify items in the world. Spoken and written terms signify concepts and only derivatively signify extra-mental objects. The ultimate significate of any term is a concrete, individual item or a set of such items. Any significative term that can signify something extra-mental does so thanks to concepts. BOB also embraces tokenism about all linguistic and mental items.

This basic account of signification grounds accounts of supposition (*suppositio*). At least since the thirteenth century, supposition had become the pivotal doctrine in medieval semantics.[24] Supposition is the property and the function of terms in propositions to "stand for" or "pick out" the objects they signify. While signification can occur with isolated, free-floating terms, supposition is exclusively a property of terms in sentences.[25]

What follows focuses exclusively on personal supposition, which is roughly equivalent to what we call reference and involves a term's standing for objects in such a way that the proposition can be verified through demonstrative reference. A distinctive feature of BOB's semantic theory is this demonstrative model of reference that Buridan, Marsilius, Albert and their fellow mid- to late- fourteenth-century

---

[21] On this, cf. Ciola (2020).

[22] Cf. Parsons (2014). Our BOB is a next generation, more overtly nominalistic offspring of Terry's BOB that included also bits and pieces from other authors, as, e.g., Walter Burley.

[23] Medieval semiotics and semantics have been in the past few decades a fast expanding area of research. While somewhat dated, a clear and insightful introduction to these topics remains Spade (1996).

[24] Supposition theory is to this date one of the most studied theories in medieval logic and semantics. For helpful and relatively recent introductions on the subject, see Kann (2016), Parsons (2008).

[25] Cf. de Rijk (1971), (1973, esp. 37ff).

nominalists loosely inherited from Ockham.[26] The truth conditions for statements like "Brownie runs" (*Brunellus currit*) can be analysed as "this is Brownie and this runs" (*hoc est Bunellus et hoc currit*), where the demonstrative pronoun "this" points to some present object. This is effectively a pointing-criterion (*additamentum*) for personal supposition, which is somewhat typical of fourteenth-century nominalism. Terms have supposition when there is some *thing*—a *suppositum* or multiple *supposita*—that can be pointed at by this demonstrative pronoun. When there is no such a *suppositum*, the terms are empty; if there cannot be any such a *suppositum*, the terms are *necessarily* empty.

This demonstrative model reflects several theoretical commitments since it primarily aims to ground reference in an ostensive and empirically based procedure with the intent to be epistemologically transparent and ontologically grounded. However, basing one's articulation of supposition on a pointing-criterion reinforces the ties that articulated theories of supposition in general already entertain with present existence, since pointing at something requires that there is in some sense some thing to be pointed at. So, the demonstrative model also creates immediate problems for handling objects that are not presently available for ostensive reference. For if reference requires the possibility of demonstrative identification, then statements about past objects, future objects, merely possible objects, and impossible objects all become problematic. The nominalist pointing-criterion underlying personal supposition thus highlights the problem of empty terms as a systematic core challenge rather than a mere oddity in limit-cases.

Problems with empty terms are far from uncommon in late-medieval logic. Trivially, they immediately arise from Aristotle's account of the different types of categorical sentences and the relationships subsisting among them; these are the logical relationships that support the basic logical operations articulating Aristotle's syllogistics.[27] Categorical sentences may be either affirmative or negative, and at the same time, they may be either universal or particular with respect to their quantity. This yields four kinds of categorical sentences: universal affirmative sentences of the form SaP, e.g., "Every human being is mortal"; universal negative sentences of the form SeP, e.g., "No human being is mortal"; particular affirmative sentences of the form SiP, e.g., "Some human being is mortal"; and particular negative sentences of the form SoP, e.g., "Some human being is not mortal." For the logical and semantical relationships of the Aristotelian square of opposition to subsist, SaP sentences entail by subordination SiP.[28] But particular affirmative sentences are read as existential statements in their premodern understanding as well as in the contemporary one. For instance, "Some human being is mortal" amounts to something like "There is a human being and that human being is mortal." So, if SaP is true, SiP must be true

[26] On Ockham's "pointing criterion," see Spade, Panaccio, and Pelletier (2025), especially §3.2: "Supposition and Signification."

[27] Read (2015).

[28] Cf. Parsons and Ciola (2025).

as well. But this implies that SaP cannot be vacuously true, which is contrary to its standard reading in contemporary classical logic.[29]

In other words, affirmative statements can be true if and only if their subject terms have *supposita*. This is the assumption of Existential Import for affirmative sentences. If the subject of an affirmative statement has *supposita* and these *supposita* are as the predicate describes them to be, then the affirmative statement is actually true. If the subject does not have *supposita*—or if has *supposita* but they are not as the predicate describes them to be—then the affirmative sentence is false.

Negative sentences, by contrast, can be true either when the subject fails to be as the predicate describes or when the subject term is entirely empty. In other words, negative sentences do not have Existential Import. For example, "No student is in the classroom" is true either if there are students enrolled in the course but none of them made it to class, or if there is no student at all to begin with. This is the case as well for SoP sentences, which might sound counterintuitive with our normal understanding of ordinary language. For instance, "Some students are not in the classroom" will be true also if no student at all is in the classroom.

Besides the need to preserve the logical relations of the Aristotelian square of opposition, the temporal connotations built into the copula "is" (*est*) reinforce affirmative sentences' existential commitment by importing present-tense reference.[30] This seems to be mid-fourteenth-century Parisian Nominalism's baseline understanding of the copula's importing the present when in its present tense form, which is the norm in the standard treatment of reference in the context of supposition theory.

Then, it seems almost unavoidable for these commitments to entail that affirmative categorical statements with presently empty subjects are false by default. But in some cases, this is an obviously undesirable side effect as at least three important classes of statements would be false by default.[31] These are the following.

A. Statements about presently empty natural classes, such as:

"Roses are flowers" when stated in winter and there are no flowers;

"Every lunar eclipse is for the interposition of the Earth between the Sun and the Moon"[32];

---

[29] In contemporary classical logic, the true logical form of a SaP sentence is understood to be that of a conditional stating, for example, that for every x, if x is a human being, then x is mortal. In case there are no human beings and thus x is not a human being, the antecedent would be false; consequently, the whole conditional would be "vacuously" true by default.

[30] A common medieval formulation is in terms of the copula cosignifying the present. See Maierù (1972). See, e.g., Buridanus n.d., *Summulae de Dialectica*, 1.2 (transcr. Hubien): "copula semper consignificat certum tempus simpliciter determinatum ad praesens, praeteritum vel futurum, participium autem vel verbum infinitivi modi, licet consignificet tempus, tamen hoc non est simpliciter determinate ad praesens, praeteritum vel futurum, sed respective ad copulam."

[31] Recently, I addressed these matters more extensively in Parsons and Ciola (2025), especially §5. Cf. also Ciola (2019) and (2020). I relay on my expositions in these studies for the section that follows.

[32] E.g., Buridan discusses this example (*Omnis eclipsis lunae est propter interpositionem terrae inter solem et lunam*) at length in SDD, (1998, 4.3.4, 47).

"Thunder is sound made in clouds" when stated on a clear day[33];
"Every rainbow is a reflection or refraction of light"[34];
"Dodos are birds" as dodos are extinct.

B. Statements about idealised or abstract items impossible in nature despite not being intrinsically contradictory, such as:

"The void is place without matter," which is impossible in the broadly Aristotelian physical understanding of the natural world, which was the commonplace received view in the later Middle Ages;
"Idealised gases are infinitely compressible," which is naturally impossible but is a useful abstraction to model the physical behaviour of real gases;
"A point is space with no extension," which expresses a pure mathematical abstraction.

C. Even trivial identities with necessarily empty subjects, such as:

"The round square is the round square"[35]
"A chimera is a chimera"—where the chimera is the medieval counterpart of the round square—i.e., an intrinsically contradictory item constituted of incompossible parts.

These three classes collect statements about items with diverse ontological standings. A-statements involve terms that could have present referents but happen not to: i.e., they refer to naturally possible kinds that are temporarily or permanently absent—but this was not or will not always be the case. B-statements involve relative impossibilities, that is, entities that cannot exist given the laws of nature but are not intrinsically contradictory. Instead, C-statements involve absolute impossibilities that violate logical or metaphysical principles like, e.g., the law of non-contradiction. But under the assumption of Existential Import, all statements belonging to these groups are equally false by default.

Maybe, renouncing the truth of some identity statements might not be too high of a price to pay; most late-medieval logicians subscribe to this claim, including Buridan[36] and Albert of Saxony.[37] But the default falsity of A- and B-statements has serious implications on our ability to state affirmative scientific truths. Thus, this default falsity poses a serious challenge for nominalist semantics, for many statements that seem obviously true—scientific generalisations, mathematical theorems, logical principles—involve subjects that lack present referents. If nominalists cannot account for the truth of such statements, their semantic theories will be systematically inadequate to model scientific and mathematical argumentation even by their own scientific standards.

---

[33] E.g., Buridan discusses the example *omne tonitruum est sonus factus in nubibus* in *Quaestiones super decem libros Ethicorum Aristotelis ad* Nicomachum, 1513, VI, q. 6, f. 118r-119v, and famously in *Summulae de Dialectica* (henceforth SDD), (1998, 4.3.4, 46–47).

[34] E. g., Buridan discusses the example *omnis iris est reflectio aut refractio lucis* in SDD, (1998, 4.3.4, 46).

[35] In contemporary philosophy, the Round Square has been popularised by Quine 1948 and is the standard example of intrinsically contradictory item.

[36] E.g., Buridanus (1976, 1.5.3).

[37] E.g., Albert of Saxony, *Questions on Logic* (2002, q. 13).

The problem cannot be solved by exceptionally rejecting existential import and simply analysing away SaP sentences as vacuously true conditionals when we operate in a scientific context. This may indeed seem an obvious workaround, and the idea was clearly circulating already in the first half of the fourteenth century since William of Ockham discussed it. But Ockham also determined that it would not solve much and was, thus, ultimately untenable. This is because late-medieval scientific argumentation finds its standard expression in the type of demonstrative syllogistic presented in Aristotles' *Posterior Analytics*, which does not run on conditional premises but on premises that are categorical, affirmative, universal and necessarily true.[38]

Alternatively, letting go of the demonstrative model of reference would undermine the empiricist and ontologically parsimonious approach that inspired and characterised nominalist semantics in the first place.

The challenge is, therefore, to develop strategies for handling empty terms that preserve the advantages of nominalist semantics while accommodating the truth of problematic statements. The solutions developed by medieval nominalists represent sophisticated—even if not always coherent—attempts to eat their cake and have it too.

## 5.4 Sampling Semantic Strategies: Natural Supposition vs. Ampliation

### *5.4.1 Thunder and Roses: Natural Supposition and Its Limitations*

Buridan revives and reinterprets a theory that was widespread at an earlier developmental stage of late-medieval semantics but had since then fallen out of fashion, namely natural supposition (*suppositio naturalis*).[39] In Buridan's reconceptualisation, natural supposition becomes a kind of omnitemporal or atemporal supposition.[40] A term has natural supposition when it "supposits indifferently for all the

[38] Cf. William of Ockham, *Summa Logicae*, 1972, III–2.5, ll. 46–47. Note that de Rijk (1973, 52) mistakenly believed Ockham to endorse such a conditionalisation workaround. The passage is discussed in more detail in Parsons and Ciola (2025).

[39] Cf. de Rijk (1971, 1973), Maierù (1972). Buridan himself remarks on the lack of endorsement or more generally interest for natural supposition among his own contemporaries—cf. Buridanus (1998, 4.3.3, 45): "Multi moderni negant suppositionem naturalem ad istum sensum qui positus est in textu. Sed sine recitatione opinionum ego ostendo quod saepe utumur ea."

[40] There is a pending discussion in the scholarship to assess whether Buridan's account of natural supposition should be understood omnitemporally—i.e., as supposition picking out referents in all times—or atemporally, i.e., as a kind of supposition removed from time and tense connotations. Besides De Rijk's two-part study cited above, see also Fitzgerald (2006). Buridan's extended examination of natural supposition in SDD, (1998, 4.3.4) supports both readings, and it might be the case that the whole issue has to be radically reframed and reassessed in future studies.

things for which it can supposit, equally for present and for past or future things."[41] This temporal expansion partly removes supposition's exclusive connection to the present tense. This amounts to reading the copula "is" (*est*) as a predication that is not narrowly restricted to the present tense but is instead some kind of omnitemporal or atemporal predication. Through this special reading of the copula, statements about presently absent objects can be true in virtue of referents that may not actually be there.

For Buridan, natural supposition is already commonly employed in scientific statements, especially in those instances where general statements about natural classes should still be true independently on whether those natural classes have any actual instantiation at present.[42] When we define thunder as "thunder is sound made in clouds," we are not making a statement that can be verified only by actual thunder roaring in the sky at the time of our speaking. We are making instead a statement that is true and must be true whenever we state it about any thunder—be it past, present, or future. The truth of this general definition must be independent of any thundering happening at the moment of enunciation—provided that there have been or there will be instances of thunder that were and will be sounds made in the clouds. Thus, natural supposition helps defuse the issue of empty reference under special circumstances, namely in scientific, demonstrative statements about natural classes that may happen to be empty.

But Buridan maintains that natural supposition is also triggered by intentional verbs—i.e., in the medieval jargon, "verbs signifying an act of the intellective soul," like "understanding," "knowing," "cognising," "opining," and "signifying."[43] For Buridan, "these verbs make the term constructed with them supposit or be taken for present, past and future things."[44] Then, "I understand roses" can be true even in winter when no roses are present: the understanding relates to roses generally, not just to presently existing roses, and to my mind's entertaining of the concept of "rose" in the specific epistemic attitude of "understanding." Yet, Buridan's claim about intentional verbs is unrestricted: for all intents and purposes, they would import natural supposition also when I am not understanding, knowing, signifying, etc. roses, thunder, dodos and the like, but anything that can be understood, known, signified and so on.

Buridan also appeals to natural supposition for explicit temporal generalisations like "everything that is or was or will be runs." Here, Buridan claims, the complex subject ("everything that is or was or will be") supposits for referents across different times. Since supposition is a property of a term in sentences and ultimately depends

---

[41] Buridanus (1998, 4.3.4, 45): "Naturalis suppositio vocatur secundum quam terminus indifferenter supponit pro omnibus pro quibus potest supponere tam praesentibus quam praeteritis vel futuris."

[42] Buridanus (1998, 4.3.4, 45): "Et hac suppositione utimur in scientiis demonstrativis."

[43] Buridanus (1998, 4.6.1, 90).

[44] Buridanus (1998, 4.3.4, 45): "Primo quia ista verba 'intellegere,' 'scire,' 'cognoscere,' 'opinari,' 'significare,' et alia multa faciunt supponere vel teneri terminos cum eis constructos pro praesentibus, praeteritis et futuris. Unde sequitur 'homo qui fuit, intellegitur (ut Aristoteles); ergo homo intellegitur.' Similiter 'homo qui erit, intellegitur (ut Antichristus); ergo homo intellegitur,' sicut ego dixi in *Sophismatibus* capitulo quinto *De ampliationibus*."

on the copula, the complex subject containing temporal features seems to affect the reading of the copula in a sentence such as "everything that is or was or will be is running."[45] Then, the signification of the complex subject makes the copula omnitemporal or atemporal; and the omnitemporal or atemporal copula makes the subject have natural supposition. Buridan extends the scope of natural supposition to other cases and offers an extensive analysis in SDD (1998) 4.3.4.[46]

Natural supposition solves many Class-A problems but is not fully satisfactory. Buridan's account seems to restrict natural supposition to entities that actually exist at some point in time, seemingly excluding purely possible entities never actualised. This would exclude pure *possibilia* like the golden mountain,[47] creating problems for scientific and mathematical statements involving theoretical entities like dimensionless points or the void. Therefore, strictly speaking, no affirmative statement about such entities could ever be true if natural supposition cannot apply to them—assuming that natural supposition is the only available theory in Buridan's toolbox.

Second, there is a systematic tension between Buridan's treatments of natural supposition and ampliation. Buridan lists intentional verbs among the causes of natural supposition, but he also treats intentional verbs as paradigmatic causes of ampliation. Since ampliation is explicitly designed to extend personal supposition beyond present existence, this creates an overlap and a theoretical redundancy.

The problem becomes evident in the *Tractatus de Consequentiis*. Here, Buridan states that intentional verbs—i.e., verbs such as the above-mentioned "understanding," "knowing," "cognising," "opining," and "signifying"—cause ampliation to purely possible objects, not just to past and future actual objects. But this directly conflicts with Buridan's account of natural supposition in terms of temporal expansion across present, past and future existents, which would be imported by the very same verbs. If intentional verbs can reach purely possible referents through ampliation, it is unclear why the natural supposition imported by the same verbs should be restricted only to past, present and future referents. As is unclear whether the same verbs could alternatively be causes of ampliation and causes of natural supposition—and, in that case, what would determine when they would cause the one rather than the other.

### 5.4.2 The Quest for the Purely Possible (and Beyond): Ampliation and Its Broadening Alternatives

Buridan's temporal characterisation of natural supposition apparently excludes *pure possibilia* from being viable referents. Yet, many philosophical and scientific contexts

[45] This is but the canonical logical form of "everything that is or was or will be runs."

[46] The whole section SDD (1998, 4.3.4, 4–49) is devoted to natural supposition.

[47] It should be noted that, in the fifteenth and sixteenth century logical tradition, cases such as the golden mountain become paradigms of absolute impossibility, as John Mair exemplifies. See Alt (2025).

seem to require reference to purely possible entities to verify their generalised affirmative statements. If natural supposition cannot handle subjects referring to mathematical items, nominalist semantics remains stuck between a rock and a hard place. This limitation is only the more evident in intentional contexts since they often seem to involve purely possible objects. I positively and affirmatively understand that mathematical points are dimensionless; analogously, I can think of a possible golden mountain as being such and such. If natural supposition cannot save these thoughts from default falsity, it fails an important modelling task. These persisting problems and theoretical constraints partly explain why natural supposition did not make a wide or perduring comeback in the fourteenth century despite Buridan's revamping of the theory. Instead, many of Buridan's contemporaries and followers deployed theories of ampliation as an alternative approach to these and other issues.

In its broadest characterisation, ampliation is one of the ways in which late-medieval semantics handles tensed and modal referential contexts. For handling empty terms, ampliation offers several advantages over natural supposition. For instance, it provides a unified framework for handling temporal, modal, and intentional contexts, thus yielding a unified treatment of both problematic scientific statements and ordinary language.[48] Furthermore, it can easily accommodate purely possible objects. And finally, some versions of ampliation can integrate more smoothly the pointing-model of reference characterising nominalist semantics by *expanding* on it rather than creating ad hoc exceptions. Ampliation does so by admitting different types of extension beyond present existence, which allows for more a fine-grained treatment of different kinds of empty terms. This flexibility proves useful for handling the full range of problematic cases. Overall, ampliation is a more powerful theory than natural supposition, plain and simple. It is actually a matter of discussion whether it actually does a better job than natural supposition and is indeed compatible with the tenets of "nominalist semantics."[49]

The basic idea is simple: ampliation is an expansion of supposition that allows terms to supposit for items beyond those available in standard present-tense contexts. As John Dorp of Leiden[50] explained around 1393 in his influential handbook, "ampliation is wide[r] supposition" that extends the domain of reference in different ways.[51]

Early applications focused on cases like "A black thing will be white," requiring reference to objects different from their present state.[52] By the fourteenth century, ampliation had become a rigid theoretical genre with systematic rules for identifying ampliative contexts and determining the scope of ampliation. The theory was not limited to specific cases but had developed into a general framework for

---

[48] I treated this point more extensively in Ciola (2019).

[49] A sceptic view has recently been expressed by Pasnau (2025). See also Pasnau (2020).

[50] On John Dorp of Leiden and the major influence on his commentary, see Bos (2000), Wojtczak (2000).

[51] Dorp 1487, no foliation, Tractatus IV: *De suppositionibus*: "ampliatio est ampla suppositio."

[52] Parsons (2008).

understanding the relationship between reference and different temporal and modal domains.

Ampliative sentences contain markers including: tensed copulas, modal copulas and terms, participles, adjectives in *-bilis* (e.g., *intelligibilis*), and intentional verbs (e.g., "understanding," "knowing"). These markers are morphological or syntactic indicators that ampliation occurs, providing a general criterion and moving beyond ad hoc treatments of individual special cases—as were the special contexts handled by natural supposition.

Theories radically differ on how ampliation is articulated, how far it extends, and what kinds of objects it can reach. A clear overview occurs in John Dorp of Leiden's widely circulating semantics handbook. This work is presented as a commentary on John Buridan's *Summulae de Dialectica*, but Dorp often digresses and develops extensive discussions that are largely independent of Buridan's text, resulting in a largely original work. Treatise 4, on *Supposition* and related properties, does not comment on Buridan's SDD (1998), 4, but rather on a version of Marsilius' treatises on supposition, ampliation, appellation, restriction and alienation[53] that is interpolated with Buridan's SDD (1998), 4.[54] Thus, Dorp's commentary is especially representative of the dynamic evolution of the intellectual debates within "nominalist" circles around the end of the fourteenth century. It is also indicative of the direction in which these discussions will develop during the fifteenth century, and—given its extraordinary success and influence—it most likely contributes to shaping them.

Dorp identifies two fundamental ways in which supposition can be ampliated: (i) with respect to *supposita* or referents, and (ii) with respect to time or tense (*tempus*).[55]

At face value, this distinction echoes a similar division proposed by Peter of Spain in the *Tractatus,* the seminal text that sets the starting point for these discussions in the so-called tradition *Summularum.*[56] But the criterion underlying Peter's division was not entirely clear. Both Peter's examples addressed modal cases. On the one hand, ampliation over *supposita* seemed to treat sentences involving possibility, as in "Human being[57] can be the Antichrist." On the other hand, Peter's account of ampliation with respect to time or tense seemed to handle sentences expressing some kind of necessity, as in "Human being is necessarily an animal." Depending

[53] These treatises are edited by Bos (1983).

[54] Bos (2000) had already remarked that Marsilius' text was the basis for this section of Dorp's commentary.

[55] Dorp 1487, no foliation: "Pro quo notandum est secundo quod terminus dicitur dupliciter ampliari. Uno modo quoad supposita. Alio modo quoad tempus." Latin does not distinguish between "time" and "tense"—the grammatical feature of the verb expressing time—unlike English. I will use the terms interchangeably unless otherwise specified.

[56] Petrus Hispanus (1972)—cf. especially Treatise 9: *De ampliationibus*, 194–195: "Et ideo sequitur alia divisio ampliationis, scilicet quod ampliationis alia est respectu suppositorum, ut 'homo potest esse Antichristus'; alia respectu temporis, ut 'homo necessario est animal,' ut dictum est."

[57] Here I maintain the indefinite quantity of the Latin sentence in the English rendering and the resulting ambiguity. Medieval logicians tend to read these unspecified quantifications either as universals (i.e., "every human") or as particulars (i.e., "some human").

on one's underlying understanding of modalities,[58] Peter's division could be read in different ways but remains unarticulated and unclear.

While echoing Peter of Spain's phrasing, Dorp's distinction captures two fundamentally different ways of conceiving ampliation's object, functioning, and scope. Dorp briefly articulates both views.

(i) Ampliation with respect to *supposita* expands the number of referents. For example, in "Human being will be white," the term "human being" supposits for more humans than it would in a present-tense context because the reference of the ampliated term includes future humans who do not yet exist.[59]

This conception of ampliation is the most widespread and traditionally consolidated. Peter of Spain himself must have endorsed a view of this kind since he considered personal supposition itself a property of common terms—so, a fortiori, the same must be the case for ampliation.[60]

Even in the fourteenth century, the limitation of ampliation to common terms persists as a side effect of conceiving ampliation in terms of "more referents." Ampliation so characterised can only apply to common terms taken commonly—that is, to terms that are not discreet and thus have more than one referent they can pick out even at the present tense. I can ampliatively say something like "Some dogs will be white" as they grow old, but I cannot ampliatively say "Marlowe will be white." Of course, my dog Marlowe will be white as he grows old. But ampliation with respect to *supposita* cannot treat this or any other discrete statement.

Buridan and Albert of Saxony endorse versions of this theory but interpret ampliation's starting point differently. Buridan's starting point is the present expressed by a term's *status*, i.e., the exact referents for which the term supposits in the present-tense context. To this *status*, Buridan's ampliation adds additional objects from the temporal or modal period indicated by the tense of the copula or by its modal modification through an ampliative term. Albert is a stronger actualist: his reading of the present is not simply in terms of the time expressed by the present-tensed copula, but in the ontologically stronger terms of existing actuality.[61] Albert's ampliation starts from items that actually exist and expands to include past, future and even possible items. This different conceptual and ontological emphasis does not affect the technical articulation of their theories in purely tensed contexts, but it results in a different treatment and interpretation of modal sentences.[62]

---

58 On Peter of Spain's use of modal terms in the context of thirteenth-century modal logic, see Spruyt (1994).

59 Dorp 1487, no foliation: "Unde terminus dicitur ampliari quoad supposita qui pro pluribus supponit quam si non esset ampliatus; ut in ista: 'homo erit albus,' ly 'homo' ampliatur quoad supposita quia pro pluribus supponit quam supponeret si non esset ampliatus. Nam supponit [pro] omnibus hominibus sive presentibus sive futuris."

60 Petrus Hispanus (1972, 194): "Personalis suppositio est *acceptio termini communis* pro suis inferioribus." (The highlight is mine).

61 Cf. Fitzgerald (2006, 315–24).

62 In particular, Albert's treatment of contexts ampliated to the possible has some problematic consequences already remarked by some of his quasi-contemporaries, including Marsilius in his

Despite differences of articulation and interpretation, all such theories share the strategy of expanding the number of *supposita* and implement ampliation proper on common terms only.

(ii) Ampliation with respect to time involves intervening on the domain of reference itself.

Paraphrasing Dorp: a term is ampliated with respect to time when it supposits for its significate or its significates with respect to different times indifferently. In other words, this ampliation takes place with respect to different copulas importing other tenses or modal spaces, and it does so with respect to any of them disjunctively—as is customary in ampliative analyses.[63]

Effectively, this alternative characterisation of ampliation expands the tensed or modal domains within which supposition can occur, treating uniformly any kind of item that may occur in any of those domains as long as the ampliative analysis scopes over more than one tensed or modal domain. As Dorp himself noticed, ampliation with respect to time in some cases yields the same analysis as ampliation with respect to the *supposita.* Common terms ampliated in purely tensed contexts about the past or the future are the obvious example of such convergence: both types of ampliation can handle a sentence such as "human being will be white" and do so in the exact same way.[64] But Dorp's claim is stronger than this: whatever ampliation with respect to *supposita* can handle, ampliation with respect to time can handle too and just as well.[65] But not vice versa, for ampliation with respect to time can treat cases that fall beyond the scope of ampliation with respect to *supposita*—i.e., it is a more powerful theory. For instance, ampliation with respect to time can ampliate also discrete terms.[66]

The reminder of Dorp's examination focuses on this type ampliation since it is systematised in Marsilius' treatises upon which Dorp is commenting.[67] Marsilius' theory of ampliation is probably the most influential, fully articulated account of ampliation with respect to time, and it expands the domain of the viable "distinctions of time" (*differentiae temporum*) over which ampliation can scope.[68]

---

*Consequentiae* (2017). See, e.g., Ciola (2018b). See also Marsilius de Inghen (1983), *Ampliationes* 108.

63 Dorp 1487, no foliation: "Sed terminus dicitur ampliari quoad tempus qui supponit pro suo significato vel suis significatis respectu diversorum temporum indifferenter, i.e., respectu diversarum copularum differentias temporum importantium et respectu cuiuslibet divisim."

64 Dorp 1487, no foliation: "Patet, quia in ista 'homo erit albus' ly 'homo' ampliatur quoad supposita. Similiter ampliatur quoad tempus, quia supponit pro suis significatis in ordine ad diversas copulas differentias temporum importantes."

65 Dorp 1487, no foliation: "Ex isto sequitur quod quandunque eadem est ampliatio termini quoad supposita et quoad tempus."

66 Dorp 1487, no foliation: "Secundo sequitur quod termino singulari convenit ampliatio quoad tempus. Patet, quia terminus singularis potest supponere in ordine ad diversas copulas differentias temporum importantes. Ut in ista 'Sortes erit homo' ly 'Sortes' supponit pro illo quod est Sortes [vel] pro illo quod erit Sortes in ordine ad istam copulam disiunctivam 'est vel erit.'"

67 Dorp 1487, no foliation: "Tertio sequitur quod diffinitio ampliationis auctoris in littera solum intelligat de ampliatione termini in ordine ad tempus."

68 Cf. Marsilius de Inghen (1983), *Ampliationes*, 98ff.

## 5.5 Beyond the Domain of Possibility: Marsilius' Ampliation to *Imaginabilia* and Its Status

Marsilius' treatment of ampliation is distinctive. Beyond traditional tenses and possibility, he adds the domain of imaginability (*imaginabile*).[69] This new modal space hosts items that can be conceived even though they cannot possibly exist, and it is separate and distinct from that of possibility. This last aspect is clear, for example, from Marislius' analyses of two types of sentences, such as respectively: "the void is understandable" (*vacuum intelligitur*) and "for-man-to-be-an-ass can be signified" (*hominem esse asinum potest significari*).[70]

While the void is naturally impossible in the commonly received version of Aristotelian physics, it is nevertheless not intrinsically contradictory, and thus, it is logically possible. Therefore, the subject of this sentence finds some kind of "referent" within the domain of possibility. But the void is also understandable as such. And since "the intellect extends itself to the imaginables (*imaginabilia*),"[71] the subject term "void" finds an analogous kind of "referent" within the domain of the imaginable as well. Thus, the sentence satisfies ampliation's requirement to scope over more than one "distinction of time" (*differentia temporum*). So, the sentence is ampliative, and in this case it happens to be ampliatively true as well.

Furthermore, sentences involving absolute impossibilities make it clear that this space of imaginability is not a mere reduplication of the domain of pure possibility since it also includes absolutely impossible items. Certainly, sentences with subjects signifying intrinsically contradictory items—such as the chimera or the ass-man—cannot be ampliatively true, since chimeras and ass-men are only hosted in the domain of the imaginable but in no other. Therefore, while there is a sense in which chimeras, ass-men, unicorns and all such absolutely impossible items may be understandable or significable, sentences such as "the chimera is understandable" or "for-man-to-be-an-ass can be signified" cannot be verified through ampliation theory.[72]

---

[69] Marsilius de Inghen (1983), *Ampliationes*, 102: "Secundo circa dictam descriptionem superius positam est notandum quod tres sunt differentie temporum, scilicet presens, preteritum et futurum. Et tales etiam dicuntur tempora, ut solemus dicere 'hoc est presens tempus,' 'hoc est preteritum,' etc. Et possunt addi duo, scilicet *posse* et *imaginari esse*, que, licet, non sint proprie differentie temporum, tamen in proposito sunt differentie temporum, nam respectu eorum termini supponunt in propositionibus pro diversis temporibus, ut postea dicetur. Est igitur descriptio superius posita intelligenda sic, quod ampliatio est suppositio termini pro suis significatis sibi correspondentibus pro distinctis temporum differentiis; puta quod est suppositio termini pro suis significatis que sunt vel fuerunt vel pro hiis que possunt esse vel imaginari esse."

[70] Marsilius examines extensively the example of the void as a pivotal argument in his treatise on ampliation. See Marsilius de Inghen (1983), *Ampliationes*, 102–110. The case of the ass-man is instead extensively discussed in the treatise on *Consequentiae.*

[71] Marsilius de Inghen, *Consequentiae* (2017), I: "Intellectus enim extendit se ad imaginabilia." For an extended examination of this passage, see Ciola (2020, esp. 161ff).

[72] These points are further developed in: Ciola (2019) and (2020).

This treatment appears consistently and more or less explicitly across Marsilius' logical works. By the late fourteenth century, John Dorp emphasises the identification of imaginability with impossibility, establishing the received interpretation of Marsilian semantics that will perdure till at least the sixteenth century.[73]

For example, Dorp clearly emphasises how intentional verbs do not simply ampliate to what is in some sense possible but push straight into the space of the impossible. In his text the equation between the imaginable and the impossible is explicit: "the ampliative strength of these verbs scopes beyond the possibles to the imaginables or impossibles [...] the impossible can be imagined and signified."[74]

This reading unambiguously endorses a kind of referential treatment for impossible objects in appropriate contexts, enabling true statements about items that could never exist under any circumstances and may even be contradictions.

This account provides unified treatment of empty subjects in scientific and ordinary language, verifying Class-A and Class-B statements. But Marsilius' treatment fundamentally rewires reference, enabling a full semantic treatment of absolutely impossible items. Though "the chimera is imaginable" (*chimera est imaginabilis*) cannot be ampliatively true (chimeras occupy no temporal or modal domain except imaginability), occupying one modal space is significant. For occupying one modal space, namely that of the imaginable, is better than nothing, and there is a sense in which *the chimera is imaginable.* This is a non-trivial sense and relies on a twofold fundamental shift in Marsilius' basic understanding of supposition and ampliation theory: (a) bypassing present reference in ampliation by (b) deploying a more primitive notion of reference expressed in his reconceptualisation of a term's *status.*

If ampliation expands supposition and supposition remains tied to the present, most theories remain tied to the present. But Marsilius' ampliation bypasses present reference: "the void is understood" is ampliatively true despite the fact that the void is not and cannot ever be presently instantiated. This analysis represents a significant departure from the standard treatment ampliation, especially in a nominalistic framework.

One can legitimately wonder whether Marsilius' theory is coherent, for he seems to maintain that ampliation *is* a wider supposition[75]; and supposition is grounded in the present through the pointing-criterion typical of nominalist semantics. How can Marsilius' treatment of "the void is understood" be even remotely coherent with his semantic tenets? Taking supposition proper as usually defined and articulated in treatises of supposition—even Marsilius' own[76]—as the starting point, this does not seem possible. I suggest a charitable interpretation: supposition proper doesn't exhaust the referential function.

---

[73] Cf. Ashworth (1977), Alt (2025).

[74] Dorp 1487, no foliation: "[...] [v]is ampliativa dictorum verborum se extendat ultra possibilia ad ymaginabilia aut impossibilia. [...] Respondetur quod vis ampliativa dictorum verborum se extendit ultra possibilia ad ymaginabilia et quod impossibile potest ymaginari et signifcari."

[75] Cf. Marsilius de Inghen (1983), *Ampliationes*, 108: "Ad istud dico quod ampliatio debet termino attribui respectu suppositionis et nullius alterius, et hoc videtur antea aliqualiter fore probatum. [...] ampliatio est magis ampla suppositio."

[76] Cf. Marsilius de Inghen (1983), *Suppositiones*, 52–96.

**Table 5.1** Buridan and Marsilius on *status*

| Buridan, *Summulae de Dialectica*, 4.6.1 | Marsilius, *Restrictiones et Alienationes* |
|---|---|
| First, we should consider the status according to which a term is taken as neither ampliated nor restricted, with respect to which status sometimes there is ampliation sometimes restriction. That status can be assigned when a term supposits or appellates exactly for all its significates at the present tense, as if I say "human runs" or "every human runs," "human" supposes indifferently for all the humans that are at present and not for any others. | And first, we should consider the status according to which a term is neither ampliated nor restricted<br>The status beyond which a term is ampliated and below which a term is restricted is the supposition of a term for all its supposita or for its one suppositum that correspond or corresponds exactly to the term, precisely with respect to one distinction of time: either the present only, or the past only, or the future only. For example, as in "human runs," "human is animal," "Adam was," "the Antichrist will be." |

The theory of personal supposition proper expresses a specific and strongly regimented aspect of the "suppositing" or referential function of a term in a sentence, namely its "taking […] for some thing or some things for which thing or for which things the term is verified through the copula of the proposition"[77] when that copula is at the present tense. The nominalist pointing-criterion fleshing out supposition in terms of discrete demonstrative statements such as "this is Marlowe and this runs" to explain "Marlowe runs" apparently strengthens supposition's anchoring into the present. But the theory of personal supposition proper does not exhaust the suppositing or referential function, nor does it necessarily capture its most fundamental level. Marsilius does so deploying the notion of a term's *status*, a notion Buridan had already used with a similar wording but a profoundly different connotation (Table 5.1).[78] Looking at the two texts side by side is enlightening[79]:

[77] Marsilius de Inghen (1983), *Suppositiones*, 52: "suppositio est acceptio termini in propositione pro aliquo, vel pro aliquibus, de quo, vel de quibus, verificatur talis terminus mediante copula talis propositionis."

[78] See Sect. 4 above.

[79] Cf. Buridanus (1998, 4.6.1) (*De ampliationibus et restrictionibus*), 89: "Et oportet primum videre statum secundum quem terminus nec dicatur ampliatus nec restrictus, respectu cuius status aliquando dicitur ampliatio, aliquando restrictio. Status ergo ille potest assignari quando terminus praecise supponit vel appellat pro omnibus suis significatis praesentis temporis. Ut si dico 'homo currit' vel 'omnis homo currit,' indifferenter supponit 'homo' pro omnibus hominibus qui in praesenti tempore sunt et non pro aliis." The translation in the following table is mine. Marsilius de Inghen (1983), *Restrictiones et alienationes*, 160: "Et primo considerandus est status secundum quem terminus non ampliatur nec restringitur. Status enim ultra quem terminus sepe ampliatur et citra quem terminus sepe restringitur, est suppositio termini pro omnibus suis suppositis, aut omni suo supposito, sibi adequate correspondentibus, vel correspondente, respectu unius differentie temporis precise: vel presentis tantum vel preteriti solum vel futuri solum. Exemplum, ut 'homo currit,' 'homo est animal,' 'Adam fuit,' 'Antichristus erit'." The translation in the table is mine even though I refer to Bos' translation (Marsilius de Inghen 1983, 161). In the transcription above, I correct Bos' "aut omni suo suppositio" to "aut omni suo supposito" to preserve grammatical correctness and conceptual coherence.

When it comes down to explaining what the starting point of ampliation and restriction is, both Buridan and Marsilius talk about a term's *status* rather than its supposition, which appears instead in the definition of these properties. This *status* is not something else entirely, but it is the technical cornerstone supporting supposition proper and its related properties—such as ampliation and restriction. Buridan's *status* is functionally identical to supposition proper. This is not the case with Marsilius' reconceptualisation of the notion. For Marsilius, the *status* is a term's referential function with respect to one tense only, namely the tense indicated by the copula—whatever this tense may be. While Marsilius does not list examples of modally modified copulas in his short introduction on *status*, his analysis of ampliations entails that terms should have *status* in the modal space of possibility or imaginability for ampliation to be possible in cases such as "the void is understandable." Then, in a sentence such as "the void is logically possible," the subject term has *status* in the domain of possibility. Analogously, the subject terms in sentences such as "the chimera is imaginable" or "for-man-to-be-an-ass is signifiable" must have *status* in the domain of the imaginable—independently of the fact that ampliation fails.

Therefore, Marsilius effectively reshapes the foundations of supposition's referential function freeing it from the inevitability of the present tense and tying it to the exact tense indicated by the tense of the copula. But how is this shift compatible with the nominalist pointing-criterion and the demonstrative account of supposition that Marsilius still endorses? Indeed, if the referential function still needs to be explicated by pointing at something and saying "this is a chimera and this is imaginable" to ground the status of "the chimera is imaginable," then the problem is not solved at all. For this type of account and this new notion of *status* to make sense while preserving the nominalist framework, it is the very notion of "pointing" that needs to be reconsidered and freed from the present.[80] This is exactly what Marsilius does by interpreting the pointing-criterion at a more abstract level. The demonstrative account of reference is compatible with having a minimal referential function indicated by the tense of the copula if one is not actually bound to empirically pointing at past, future, possible or imaginary items in the actual world. Instead, it is the copula itself that does the pointing by connecting the term with the object or objects the term designates exactly in the space of the domain indicated by copula's own tense. If I say "Adam was the first man," there is a non-ampliative and more primitive reading that can be analysed as "this was Adam and this was the first man." I am not pointing the finger to any Adam still wandering out there, but the copula is pointing the term "Adam" to something that was Adam within the domain of past things. The same goes with copulas pointing towards a future, possible and even imaginable *status*.

Marsilius' account of *status*—and of the different tensed and modal levels of demonstrative reference it entails—allows for a uniform semantic treatment of empty terms in general. But does it also entail an ontological commitment to the reality—in a robust sense—of all kinds of temporally displaced objects, pure *possibilia* and imaginable impossibilities? This may well depend on how one conceives the semantic relationship between terms, concepts (i.e., mental terms), and whatever it is they

[80] I made a similar point on pointing in Ciola (2020, 164–166).

signify or to which they may refer. The conceptualisation of necessarily empty terms is the limit case whose consideration may be enlightening.

## 5.6 *Imaginabilia*, Complex Concepts, and Commitments

Buridan analysed terms signifying both relatively and absolutely impossible items as complex concepts assembled from simple concepts signifying real or possible things. "Chimera" signifies well but signifies no single thing out there (*bene significat, sed tamen non significat aliquid*).[81] The spoken or written word "chimera" signifies a complex concept composed of simple concepts—e.g., <goat body>, <lion head>, <snake tail>, etc.—that individually signify possible items. The Chimera's impossibility arises not from the individual components but from their incompossible combination. Similarly, "void" can be analysed as a complex concept combining <place>, <absence> and < matter>.

Bruidan's approach integrates smoothly with his broader account of mental language in general. All mental terms can be either simple—i.e., directly signifying individual entities—or complex combinations of simple concepts into phrases, propositions, and arguments.[82] The complex concepts of impossible items are therefore a natural extension of the conceptualist commitments that are built into medieval nominalism while avoiding problematic ontological commitments.

Complex concepts are mental occurrences—i.e., particular thoughts that exist in particular minds at particular times—through particular acts of compositing simple concepts that are ultimately grounded in empirical reality. Thus, such concepts are meaningful and can serve as the significate of terms like "chimera" or "ass-man," but they do not correspond to any one extramental entity.

In a sense, Bruidan's analysis of necessarily empty terms in his logical works maintains a greater ontological parsimony than theories that extend the domain of reference to include impossible items. No such item is needed within Buridan's domain, which instead maintains a straightforward empirical underpinning only to individual items out there.

However, this apparently deflationist move—despite all its advantages—comes with a price. This includes but is not limited to the hiccups within Buridan's system I outlined above. As is often the case with nominalist approaches, we must spend our ontological savings on the epistemological level.[83] Or maybe epistemological concerns might be what motivates Buridan's analysis in the first place.

In Buridan's semantics, the one-on-one correspondences of the semantic triangle become refracted and multiplied in the case of empty terms. The ultimate significates

[81] Buridan famously discusses the complex signification of 'chimera,' 'void' and other such terms in the first chapter of the *Sophismata*, in particular sophism 4, among other loci. See also, e.g., Buridanus (1976) (2015, trans. Read), 1.5.4.

[82] Cf. Buridanus (2005, 1.1.6, 16–18).

[83] This point is convincingly argued in Klima (2009).

of the spoken word and complex concept are a multiplicity of separate, individual items both in the world as well as in the domain. This results in a mismatch between the apparently simple spoken term and the multiplicity of objects in the world through the mediation of a complex mental term bridging them. The fact remains that the compounding of <place> and <absence of matter> happening in the mind cannot find a correspondence out there since there is no such thing as the void in a broadly Aristotelian world. But for Buridan, the compounding of a complex concept is functionally and ontologically equivalent or very similar to the compounding of any proposition.[84] Thus, this mismatch will be reiterated at the purely logical level. Alternatively, one might end up committing to complex significates within the domain, which Buridan—a notorious opponent of *complexe significabilia*—would definitely not want.

Theories accepting *imaginabilia* do not produce a similar mismatch. In this sense, a semantics like Marsilius' can signify and refer to abstract items subsisting as such exclusively within a slice of the domain without necessarily committing to any claims about their complexity—or lack thereof—outside of it. Functionally, Marsilius' semantics can signify the void or the chimera as one significate or one "referent" matching the one term "void" and the one term "chimera." Any considerations on their ontological structure do not pertain to Marsilius' semantical examination. Within such a theory, we can make possibly true affirmative claims about things that cannot exist, and we can do so with a certain ease without committing to much beyond our concepts. Rather than embracing robust ontological commitments, Marsilius' semantics keeps out these extralogical considerations; Buridan's, instead, does not.

Nevertheless, Buridan's ontological commitments are per se just as minimal and are more easily recognisable as such at face value. But the main concerns underpinning Buridan's complex-concept analysis of empty terms are first and foremost epistemological. For such a model explains how we can entertain the concept of a "chimera," of the "void" and of anything else of which we cannot have any direct experience—given that concepts are formed through an abstractive process from empirical perception and sensory input.[85] Buridan's account can give a relatively easy answer: we acquire the various simple concepts through the usual pathways of empirical experience, and the mind adds the composition to the mix.[86] These seem extralogical considerations that have primarily to do with our epistemic access to the world rather than to the connection between a term and the domain of reference in tensed or modal contexts.

---

[84] There is some ontological nuance here through the distinction between distant and indistant complex concepts that Buridan articulates in his *Questions on Metaphysics*. On this point, see Andrés Molero, forthcoming.

[85] For an overview of medieval theories of perception and cognition, see Pasnau (1997). For a dense analysis of some epistemic issues in Buridan and Marsilius, see Reina (2002).

[86] Of course, the story is going to be quite a bit more complicated if we want to explain where any specific way of composing incompossible concepts comes from and how it is not random, but this is a matter beyond our current scope.

The contrast is striking in comparison to Marsilius' alternative semantic approach. Marsilius' semantics—I suggest—should be read as mostly unconcerned with any considerations external to the theory; it focuses primarily on the connection between words and concepts; and it approaches the ampliative domain of reference as something more abstract. Buridan, on the contrary, seems to maintain the actual world as the domain of reference's stronghold and stresses primarily the grounding of concepts in individual, concrete objects.

But two common features are worth noting. First, in both approaches, the symmetry of the traditional semantic triangle unfolds—albeit into two different directions. Second, both approaches express poignant nominalist facets. Marsilius' semantics allows for a lack of ontological and epistemological commitments that highlights a formalist and emphatically abstract approach to logical matters.[87] Buridan's semantics puts a strong emphasis on ultimate individual referents that are causally active in normal cognitive processes—which is a theme especially dear to contemporary nominalists as well.

## 5.7 Complexity and Causation: Divides in the Tradition

Burdian's appeal to complex concepts connects his semantics to a causal genealogy of concepts, i.e., to an account of concept formation. As mentioned, medieval nominalism does not generally share contemporary nominalism's focus on causation to determine which objects are concrete and thus admissible in one's nominalist ontology. Yet, causal concerns go out by one door and come back through another. From the mid-fourteenth century onwards, causal discussions concerning empty terms, the formation and the structure of the corresponding concepts emerge more and more explicitly.

At the end of the day, Buridan's treatment of necessarily empty terms is a causal account of how we can possibly entertain concepts that do not match anything in the world and, thus, that we cannot acquire through direct experience of *one* real thing matching the concept. <Chimera> comes from our normal cognitive, abstractive process from individual instances of goat bodies, lion heads and snake tails.

However, when our semantics manipulates *imaginabilia,* it is a different story. Marsilius does not tell this story in his logical writings—probably because it is a story that does not pertain to logic proper—while his metaphysical works articulate an account that does not seem too far from Buridan's.[88] But between the late-fourteenth and the fifteenth century, the authors commenting on Marsilius' semantical texts and discussing *imaginabilia* do not refer to Marsilius' *Metaphysics* even if they engage in a discussion about the structure of concepts such as <chimera> or < void>. Yet,

[87] For instance, the ontological and epistemological underpinnings to *imaginabilia* that Marsilius examines in his *Questions on Metaphysics* seem very close, if not even identical, to Buridan's. On this point, see Andrés Molero, forthcoming.

[88] Cf. Andrés Molero, forthcoming.

at the core of this discussion, there are once again the same epistemological and ontological concerns that motivated Buridan's analysis. I shall briefly outline two representative examples: the Hagenau Commentary[89] and Dorp's *Summulae*.

### *5.7.1 Hagenau*

The Hagenau Commentary is an anonymous commentary mostly based on sections of Marsilius of Inghen's *parva logicalia* although it is presented as a commentary on Peter of Spain himself.[90] Until recently, this work has been sometimes wrongly attributed to Marsilius.[91] The commentary is best known through its Hagenau 1495 print, but Jenny Ashworth identified an earlier Basel print (1487) as well as a later Hagenau reprint (1503).[92] Despite the lack of an identifiable manuscript tradition giving material support to an earlier dating, both Ashworth and Alfonso Maierù agree that the text can likely be dated to the late-fourteenth or the early-fifteenth century on the basis of doctrinal considerations.[93] The picture of the Marsilian tradition that Hagenau outlines—as well as its substantial alignment with Dorp's treatise—supports this dating.

Examining ampliation, Hagenau contrasts Buridan's and Marsilius' positions on the treatment of terms signifying impossibilities. Hagenau focuses exclusively on absolute impossibilities—such as chimeras and ass-men—and radically contrasts these two "famous opinions" primarily on the level of signification.[94] The same contrast occurs on the level of understanding—which is coextensive to signification[95]—and it trickles down to supposition.

Hagenau reports that Buridan and his followers maintain that the impossible is not signifiable for several reasons. A crucial point is that "anything which is signifiable must truly move the intellect; but an impossible thing cannot move the

[89] Anonymous 1495; repr. Frankfurt 1967.

[90] Cf. most recently Alt (2025).

[91] This is, for example, still the case in the Frankfurt 1967 reprint.

[92] Cf. Ashworth (2013).

[93] Ashworth (2013), Maierù (1972, 234).

[94] Cf. Anonymous 1495, no foliation: "[...] utrum impossibile sit significabile. Et de isto sunt due famose opiniones auctorum Buridani et Marsilii." Hagenau is not the only commentary presenting Buridan's and Marsilius' semantics as two clearly identifiable "nominalist" alternatives especially insofar as empty terms are concerned. For example, another relevant case is the Anonymous commentator to Buridan's SDD in ms. Paris, BNF lat. 14716. Cf. de Rijk (1971) and Fitzgerald (2006).

[95] The extensionally equivalent treatment of what is understandable and what is signifiable is explicit in Marsilius' logical texts; systematic remarks in this sense occur both in the semantical treatises and in works which are not primarily concerned with the properties of the terms, such as the treatise on *Consequentiae*, 2017. Also Buridan seems to treat what is understandable as coextensive to what is signifiable. However, as Hagenau remarks as well, Buridan and Marsilius disagree on how far that extent goes. Especially on Marsilius, see Ciola (2020).

intellect; therefore it is not signifiable."[96] Impossible things cannot be understood or signified because they cannot affect the intellect through any ordinary cognitive process: impossibilities are causally inert. Hagenau's reading of the Buridanian analysis remarks that no term—be it simple or complex—can signify the impossible. A simple term such as "man" signifies only men while a simple term such as "ass" signifies only asses but no ass-men.[97] In the case of complex terms, Hagenau's main argument is mereological: complex terms "do not signify otherwise than how their parts signify; and the parts do not signify the impossible, therefore neither does the whole."[98]

On the contrary, Hagenau is explicit in attributing to Marsilius and his followers the view that terms like "chimera" signify "one item (*unam rem*) that neither is nor was nor will be nor can be but that can be imagined. Analogously, for him [i.e., Marsilius], 'man capable of braying' signifies one item (*unam rem*) that neither is a man nor an ass but an entirely distinct item (*unam rem*)."[99] Thus, for Hagenau, Marsilius and his followers hold that terms signifying absolute impossibilities actually have a *unitary significate* somewhere This significate is not possible but is nevertheless imaginable; as such, it is irreducible to a sum of the possible parts that might make it up. In other words, for Hagenau, Marsilius and the Marsilians (*Marsiliaci*) agree on accepting unitary, impossible *significates* in the domain of the imaginable. However, they disagree on the primary *significants*, i.e., on the concepts doing the signifying of such impossible yet imaginable significates.[100] The divide is over simple and complex concepts. Hagenau presents two alternative positions.

On the one hand, some Marsilians claim that an impossible significate can either be signified by a simple concept or by an absolute connotative concept and even by a complex concept.[101] Hagenau does not articulate the nuances of this seemingly broad position or set of positions. However, versions of a similar take are easily identifiable in the widespread fifteenth-century reception of Marsilius' theory of ampliation. These views reprise a deeply interconnected treatment of ampliation and appellation typical of an earlier developmental stage of supposition theory.[102] On the

[96] Anonymous 1495, no foliation: "Et tenet Buridanus quod non. Et motivum est quia omne illud quod est significabile debet vere movere intellectum; sed res impossibilis non potest movere intellectum; ergo non est significabilis."

[97] Anonymous 1495, no foliation: "Nullus terminus significat impossibile. Patet, quia nec simplex nec complexus. Non simplex quia ille terminus 'homo' significat hominem solum, et 'asinus' significat asinum solum."

[98] Anonymous 1495, no foliation: "Sed complexa non significant aliter nisi sicut sue partes; et partes non significant impossibile, ergo nec totum."

[99] Anonymous 1495, no foliation: "Alia opinio est Marsilii et suorum: quod impossibile est significabile quia res que non potest esse bene potest significari per aliquem terminum, unde ly 'chimera' secundum eum significat unam rem que nec est nec fuit nec potest esse sed que imaginatur esse. Similiter, ly 'homo rudibilis' secundum eum significat unam rem que nec est homo nec asinus sed unam rem a talibus distinctam."

[100] Anonymous 1495, no foliation: "Et sunt Marsiliaci inter se diversificati."

[101] Anonymous 1495, no foliation: "Aliqui enim dicunt quod impossibile sit significabile simplici conceptu et etiam concpetu absoluto connotativo et etiam complexo."

[102] For example, see Ciola (forthcoming).

other hand, some Marsilians claim that an impossible significate is only signifiable through a complex concept. This claim rests on a standard epistemological assumption: any simple concept is acquired (*sumptus*) either by a symmetric correspondence (*convenientia*) to some thing or by difference from it. Yet, impossible things cannot enter any such relationship of correspondence. Thus, an impossible significate is understandable only through a complex concept.[103]

Overall, all Marsilians agree that an impossible significate can be signified through a complex concept; the disagreement is whether a simple concept or a connotative concept could do the trick as well. The ground of this disagreement, as presented by the Hagenau commentary, is epistemic and has to do with concept formation, i.e., with how an impossible concept can be caused since impossible things cannot possibly be effective in the ordinary perceptive process.

### *5.7.2 Dorp*

Dorp's commentary presents a similar picture and exemplifies an eclectic approach reconciling Marsilian and Buridanian instances. Already by the end of fourteenth century, Buridan's and Marsilius' approaches to empty terms had become the main references within nominalist circles. Several commentators systematically compare these two views, sometimes considering them ultimately compatible and attempting to develop a synthesis. John Dorp's *Summulae* takes such an explicitly compatibilist approach, attempting to combine Buridanian and Marsilian elements within a coherent framework. Dorp's *Summulae IV* comments primarily on Marsilius' corresponding semantical works instead of Buridan's. The circulation of Dorp's commentary and its consequent influence were vast, making his synthetic take essential for mapping the fifteenth-century developments in nominalist semantics. In general, Dorp treats Buridan's and Marsilius' accounts as complementary strategies that address different facets of the empty terms problem. For example, Dorp embraces both natural supposition and ampliation to *imaginabilia,* seemingly understanding Marsilius to have done the same.[104]

Similarly, Dorp does not present Buridan's deployment of complex concepts and Marsilius' use of *imaginabilia* in semantics as incompatible alternatives. On the contrary, Dorp embodies the kind of Marsilian—typified by Hagenau—who deploys

---

[103] Anonymous 1495, no foliation: "Aliqui autem dicunt quod impossibile solum sit significabile conceptu complexo, quia nullus terminus incomplexus significat impossibile, quia omnis conceptus simplex vel est sumptus a convenientia vel a differentia. Nulla autem est convenientia vel differentia rei ad rem impossibilem vel rerum impossibilium ad invicem. Terminus autem complexus bene significat rem impossibilem sicut ly 'chimera' significat chimera que non est sed potest intelligi."

[104] Dorp 1487, no foliation: "Ad hoc respondetur quod sic, scilicet in suppositionem naturalem et accidentalem. Pro cuius intellectu notandum est quod differentia est inter suppositionem naturalem et accidentalem, quia terminus dicitur supponere naturaliter quando supponit pro ombibus suis significatis indifferenter, scilicet tam pro presentibus quam etiam pro futuris, vel quando supponit pro suis significatis in ordine ad copulam absoluta a connotatione temporis."

complex concepts in semantics to account for the cognitive content of impossible referents within the domain of the imaginable. For Dorp, the apparent disagreement between Buridan and Marsilius is not substantive but is often reduced to a matter of technicality. Any more fundamental divergences are easily overlooked. Thus, Dorp can read Marsilius through Buridan's lens. Doing so, Dorp can include, in a discussion of semantics proper, some epistemic matters on concept formation that were close to Buridan's heart, but that Marsilius had—maybe carefully and deliberately—left out of it.

Like Marsilius, Dorp accepts that impossibilities—even intrinsic contradictions—are signifiable and are included in the domain of ampliation in some contexts, e.g., when intentional verbs are involved.[105] Once again just like Marsilius, Dorp considers the imaginable, the signifiable and the understandable to be coextensive. But then Dorp asks himself "how can this be understandable?," which highlights the same epistemic worries shaping Buridan's semantic treatment. In line with the discussions witnessed by Hagenau, Dorp moves from the consideration that, "in understanding the impossible, the intellect is caused to understand it." But what could possibly cause this understanding? This is the core issue of the ongoing debate with which Dorp engages.[106] In response, he argues that:

> the intellect is caused by possible things to understand the impossible [...]. For possible and present things represent themselves to the intellect generating a concept in the same intellect. And the intellect composes them with each other. And since they are contradictory concepts and cannot be coherent to each other, an aggregate of those concepts signifies one impossible thing (*unum impossibile*). [...] So possibilities cause the intellect to understand the impossible accidentally, since possibilities per se cause the intellect to form concepts of themselves; once these concepts are formed, the intellect itself composes them with each other. And so it understands the impossible.[107]

Dorp's account brings back causal constraints on concept formation in the core of semantics while explaining how we can form concepts of impossible objects. Impossible concepts are not caused directly by impossible objects—which would be impossible since such objects do not exist—but indirectly by the mind's compositional operations on concepts that are directly caused by possible objects. The impossibility that

---

[105] See above Sect. 5.

[106] See, e.g., Dorp 1487, no foliation: "Dubitatur secundo utrum vis ampliativa dictorum verborum se extendat ultra possibilia ad ymaginabilia aut impossibilia. Et ista dubitatio solet moveri sub aliis verbis: Utrum impossibile sit ymaginabile sensibile vel intelligibile. Et arguitur quod non. Nam si intellectus intelligeret impossibile vel ergo ad intelligendum impossibile moveretur ab impossibili vel a possibili. Non ab impossibili quia impossibile non diffundit aliquam speciem in ipsum intellectum [...]."

[107] Dorp 1487, no foliation: "Tunc ad rationes ante oppositum concedunt quod intellectus intelligendo impossibile moveretur ad ipsum intelligendum. Et cum inquiritur unde movetur, dicitur quod intellectus movetur a possibilibus ad intelligendum impossibile [...]. Nam possibilia et presentia representant seipsa intellectui generando conceptus in ipso intellectu. Et intellectus illos ad invicem componit. Et quia illi conceptus sunt repugnantes et non potentes eisdem convenire, aggregatum ex illis conceptibus significat unum impossibile. Nam licet possibilia moveant intellectum ad intelligendum impossibile hoc est de per accidens quia possibilia movent de per se ad formandum conceptus ipsorum; quibus formatis intellectus illos componit ad invicem. Et sic intelligit impossibile."

characterises these concepts arises from logical or natural incompatibilities between their components rather than from the failure of any individual component to have possible referents. And the referent of the individual components are perceived, imagined and ultimately abstracted through the ordinary perceptive-abstractive process. Thus, Dorp's commentary seems to ultimately ground the "imaginability" characterising impossibilities on the normal perceptive function of imagination involved in ordinary cognitive processes.

Both Dorp and Hagenau show how the reception of Marsilius' semantics of impossibility puts an increasing emphasis on psychological and causal issues. While Marsilius' semantics focused primarily on semantic, abstract and technical issues—*viz.*, how supposition works, what ampliation accomplishes, which statements can be ampliatively true—later discussions shift the focus on cognitive processes and concept formation. This concern seems to have been underpinning and shaping Buridan's own treatment of empty terms albeit more implicitly in his properly semantic works.

## 5.8 Conclusions

Overall, the apparent distinction between medieval and contemporary nominalism regarding causation proves less definitive than their respective theoretical frameworks might suggest. While contemporary nominalism privileges causal efficacy as a criterion for ontological admissibility, medieval nominalism does not abandon causal considerations. Buridan's analysis of necessarily empty terms through complex concepts represents fundamentally a causal account of concept formation: there are no impossible significates or referents, but only the intellect's compositional operations upon simple concepts acquired through direct empirical contact with extramental particulars. This is a view that a contemporary empiricist ready to accept a soft form of conceptualism might find sympathetic.

The later reception of both Buridan's and Marsilius' theories, as exemplified in Dorp and the Hagenau Commentary, shows an increasingly explicit engagement with causal questions concerning how the intellect can be caused to understand the impossible. The divergence between medieval and contemporary nominalism thus concerns not the absence of causal worries in medieval philosophy but rather the level at which causal requirements occur. Within the typically conceptualist medieval framework, late-fourteenth- and fifteenth-century nominalists' concerns—and commitments—seem primarily epistemological rather than ontological.

In this study, I outlined the development and reception of a nominalist semantics of impossibilities across the late-fourteenth and fifteenth centuries. First, I broadly introduced medieval nominalism as fundamentally conceptualist rather than eliminativist. The rejection of real universals that characterises medieval nominalism differs categorically from contemporary nominalism's rejection of abstract objects in general. Operating within an Aristotelian framework in which causation functions as a relative property and space-time are not absolute, medieval nominalists were not pressed to make spatiotemporal reality and causal effectiveness their core ontological criterion.

Then, I then introduced the demonstrative model or pointing-criterion of reference that ties reference to present existence through an ostensive pointing-criterion. Against this background, nominalist semantics struggles with verifying any affirmative scientific statements concerning items that do not exist now or even that cannot exist, such as abstract entities. I proceeded to present two nominalist strategies for handling this issue, namely Buridan's and Marsilius'. Buridan revives natural supposition as omnitemporal reference, extending beyond present existence to past and future objects. However, this approach does not seem to apply to purely possible entities and produces some idiosyncratic overlaps with Buridan's treatment of intentional verbs. An alternative strategy appeals to ampliation, which is a branch of semantics expanding supposition theory to tensed and modal contexts. Theories of ampliation can be vastly different even with respect to what gets ampliated—*viz*., either the sheer number of referents (ampliation with respect to *supposita*) or the temporal-modal domains of reference taken in the semantic analysis (ampliation with respect to time). Marsilius of Inghen's theory is a version of ampliation with respect to time and constitutes a distinctive contribution in the history of modal semantics. Marsilius' systematic innovation lies in extending the domain of ampliation to include the imaginable—a modal space distinct from the possible and hosting objects that cannot possibly exist. This required a fundamental reconceptualisation of the basic referential function of terms. Rather than binding all reference to present existence through present-tensed demonstrative pointing, Marsilius permits the copula itself to point at referents within whatever temporal or modal domain it indicates. This yields a uniform semantic treatment of necessarily empty terms in scientific statements and ordinary language.

I then compared Buridan's and Marsilius' approaches. In his semantic works, Buridan analyses impossible terms as complex concepts composed of possible simple concepts, maintaining strict ontological parsimony but remains unable to verify scientific statements such as "the void is understandable." Marsilius permits unitary significates within the imaginable domain without determining their ontological structure or complexity in semantics proper and without any commitment outside of the domain itself. Both approaches are nominalistic: Buridan grounds concepts empirically; Marsilius adopts a formalist, theory-internal approach. Later Marsilians (Dorp, Hagenau) integrate causal and epistemological concerns, explaining how the intellect understands the impossible through compositional operations on concepts from possible objects. The issue of causation becomes a hot-topic within nominalist semantics in a way that parallels contemporary nominalism's emphasis on causation. But unlike contemporary nominalism's ontological emphasis, the medieval focus is epistemic. Neither Dorp nor Hagenau deny that there is a single significate or referent (*una res*) in the domain of reference for terms such as "chimera" without having to admit chimeras existing in the world. The problem is now: how do I get the concept of chimera?

To this day, Marsilius of Inghen remains recognised primarily for his contributions to natural philosophy. Yet, his innovations in semantics and logic merit equal recognition. Marsilius' reconceptualisation of reference provided late-medieval semantics with a new and adaptable conceptual tool capable of handling true affirmative statements about impossible objects without increasing one's ontological commitments.

Against this background, one can indeed reasonably be a nominalist and a "Meinongian." Even more, if one conceives logic as a metaphysically neutral tool and at the same time wants to maintain some kind of demonstrative account of reference, this might be the only reasonable thing to do.

**Acknowledgements** Funded by the European Union (ERC Starting Grant, i$^2$, n. 101078785). Views and opinions expressed are however those of the author only and do not necessarily reflect those of the European Union or the European Research Council Executive Agency. Neither the European Union nor the granting authority can be held responsible for them.

## References

### *Primary Sources*

Albert of Saxony. 2002. *Albert of Saxony's Twenty-Five Disputed Questions on Logic: A Critical Edition of His Quaestiones circa logicam*. edited by Michael J. Fitzgerald. Leiden: Brill.

Anonymous. 1495. *Commentum in primum et quartum tractatum Petri Hispani*. Hagenau. Reprinted in Frankfurt/M.: Minerva 1967.

Galileo Galilei. 1929. *Le opere*. vol. 1. Firenze: G. Barbèra.

Guillelmus de Ockham. 1974. *Opera Philosophica I: Summa Logicae*. edited by Philotheus Boehner, Gedeon Gál,a d Stephanus Brown, St. Bonaventure, NY: Editiones Instituti Franciscani Universitatis S. Bonaventurae.

Johannes Buridanus. 1513. *Quaestiones super decem libros Ethicorum Aristotelis ad Nicomachum*. Paris. Reprinted as *Super decem libros Ethicorum*. Frankfurt/M: Minerva 1968.

Johannes Buridanus. 1976. *Tractatus de consequentiis*. edited by Hubert Hubien. Philosophes Médiévaux 16. Louvain: Publications universitaires. English edition: Buridan, John. 2015. *Treatise on Consequences*. Translated by Stephen Read. New York: Fordham University Press.

Johannes Buridanus. 1998. Summulae de Suppositionibus. edited by Ria van der Lecq. Artistarium 10–4. Turnhout: Brepols.

Johannes Buridanus. 2005. *Summulae de Propositionibus*. edited by ria van der Lecq. Artistarium 10–1. Turnhout: Brepols.

Johannes Buridanus. N.d. *Summulae de dialectica*. Transcribed by Hubert Hubien. https://www.logicmuseum.com/wiki/Authors/Buridan/Summulae_de_dialectica/Liber_1/Cap2

Johannes Dorp. 1487. *Summule Buridani cum expositio preclari viri interpretis Johannes Dorp*. Lyon: Stephan Gueynard.

Marsilius of Inghen. 1983. *Treatises on the Properties of Terms: A First Critical Edition of the Suppositiones, Ampliationes, Appellationes, Restrictiones and Alienationes with Introduction, Translation, Notes and Appendices*, edited by Egbert P. Bos. Dordrecht: D. Reidel.

Marsilius of Inghen. 2017. Consequentiae. Ed. Graziana S. Ciola in *Marsilius of Inghen and the Theories of Consequences in the Late XIV Century*. PhD diss. Pisa: Scuola Normale Superiore.

Petrus Hispanus. 1972. *Tractatus Called Afterward Summule Logicales: First Critical Edition from the Manuscripts with an Introduction*, edited by L. M. de Rijk. Assen: Van Gorcum.

## Secondary Sources

Alt, Guido. 2025. Empty Reference in Sixteenth-Century Nominalism: John Mair's Case. *Open Philosophy* 8 (1): 20250070.

Ashworth, E. Jennifer. 1977. Chimeras and Imaginary Objects: A Study in the Post-Medieval Theory of Signification. *Vivarium* 15 (1): 57–79.

Ashworth, E. Jennifer. 2013. Descent and Ascent from Ockham to Domingo de Soto: An Answer to Paul Spade. In *Medieval Supposition Theory Revisited*, edited by E. P. Bos, 385–410. Leiden: Brill.

Bos, Egbert P. 2000. Die Rezeption der *Suppositiones* des Marsilius von Inghen bei Johannes Dorp (Paris) und in einem anonymen Prager *Sophistria*-Traktat (um 1400). In Philosophie und Theologie des ausgehenden Mittelalters: Marsilius von Inghen und das Denken seiner Zeit, edited by Maarten J. F. M. Hoenen and Paul J. J. M. Bakker, 213–238. Leiden: Brill.

Carnap, Rudolf. 1931. Überwindung der Metaphysik durch logische Analyse der Sprache. *Erkenntnis* 2: 219–241.

Ciola, Graziana S. 2018a. Marsilius of Inghen on *incipit* and *desinit* in *Consequentiae* II, Chapters 4–5, with an Edition of the Text. In *The Instant of Change in Medieval Philosophy and Beyond*, edited by Frédéric Goubier and Magali Roques, 170–198. Leiden: Brill.

Ciola, Graziana S. 2018b. Marsilius of Inghen on the Definition of *consequentia*. *Vivarium* 56 (3–4): 272–291.

Ciola, Graziana S. 2019. The Void and the Chimera: Marsilius of Inghen on *Imaginabilia*, *Medioevo. Rivista di storia della filosofia medievale* 44: 142–160.

Ciola, Graziana S. 2020. *Hic sunt chimaerae?* On Absolutely Impossible Significates and Referents in Mid-14th-Century Nominalist Logic. *Recherches de Théologie et Philosophie médiévales* 87 (2): 441–467.

Ciola, Graziana S. forthcoming. Half-Baked or Half-hearted? A Critical Guide to the Anonymous Question *Utrum appellatio et ampliatio sint necessarie* in Melk 958. In *Arts Quodlibetal tradition at the University of Vienna: Urban of Melk's Arts Quodlibet from 1429*, edited by Ota Pavlíček and Christopher Schabel. Turnhout: Brepols.

Ciola, Graziana S. 2026. John Buridan on the Ontology of *Consequentiae*: A Reading Guide. In *Validity Throughout History*, edited by Graziana S. Ciola, Milo Crimi, and Calvin Normore, 269–337. Munich: Philosophia Verlag.

Courtenay, William J. 1987. *Antiqui* and *Moderni* in late Medieval Thought. *Journal of the History of Ideas* 48 (1): 3–10.

Cowling, Sam, and Daniel Giberman. 2025. Nominalism in Metaphysics. In *The Stanford Encyclopedia of Philosophy*, Summer 2025 Edition, edited by Edward N. Zalta and Uri Nodelman. https://plato.stanford.edu/archives/sum2025/entries/nominalism-metaphysics/

Creath, Richard. 2023. Logical Empiricism. In *The Stanford Encyclopedia of Philosophy*, Winter 2023 Edition, edited by Edward N. Zalta and Uri Nodelman. https://plato.stanford.edu/archives/win2023/entries/logical-empiricism/

de Rijk, L. M. 1971. The Development of *Suppositio naturalis* in Medieval Logic I. *Vivarium* 9 (2): 71–107.

de Rijk, L. M. 1973. The Development of *Suppositio naturalis* in Medieval Logic II. *Vivarium* 11 (1): 43–78.

Disalle, Robert. 2006. *Understanding Space-Time: The Philosophical Development of Physics from Newton to Einstein*. Cambridge: Cambridge University Press.

Dutilh Novaes, Catarina. 2005. Buridan's *Consequentia*: Consequence and Inference Within a Token-Based Semantics. *History and Philosophy of Logic* 26 (4): 277–297.

Dutilh Novaes, Catarina. 2011. Medieval Theories of Supposition. In *Encyclopedia of Medieval Philosophy*, edited by Henrik Lagerlund, 1229–1236. Dordrecht: Springer.
Fitzgerald, Michael J. 2006. Problems with Temporality and Scientific Propositions in John Buridan and Albert of Saxony. *Vivarium* 44 (2–3): 305–337.
Guigon, Ghislain. 2019. Nominalism. In *Routledge Encyclopedia of Philosophy* Taylor and Francis. https://www.rep.routledge.com/articles/thematic/nominalism/v-2. Accessed 14 Sept 2025.
Hoenen, Maarten. 2025. Marsilius of Inghen. In *The Stanford Encyclopedia of Philosophy*, Fall 2025 Edition, edited by Edward N. Zalta and Uri Nodelman. https://plato.stanford.edu/archives/fall2025/entries/marsilius-inghen/
Kann, Christoph. 2016. Supposition and Properties of Terms. In *The Cambridge Companion to Medieval Logic*, edited by Catarina Dutilh Novaes and Stephen Read, 220–244. Cambridge: Cambridge University Press.
Klima, Gyula. 1999. Ockham's Semantics and Ontology of the Categories. In *The Cambridge Companion to Ockham*, edited by Paul Vincent Spade, 118–142. Cambridge: Cambridge University Press.
Klima, Gyula. 2004. Consequences of a Closed, Token-Based Semantics: The Case of John Buridan. *History and Philosophy of Logic* 25 (2): 95–110.
Klima, Gyula. 2009. *John Buridan. Great Medieval Thinkers*. Oxford: Oxford University Press.
Klima, Gyula. 2022. The Medieval Problem of Universals. In *The Stanford Encyclopedia of Philosophy*, Spring 2022 Edition, edited by Edward N. Zalta. https://plato.stanford.edu/archives/spr2022/entries/universals-medieval/
Maierù, Alfonso. 1972. *Terminologia logica della tarda scolastica*. Rome: Edizioni dell'Ateneo.
Andrés Molero, Gabriel. Forthcoming. From Simple Truth to Mental Complexity: John Buridan and Marsilius of Inghen on Concepts. In Truth, Falsity and Lying: Proceedings of the 24th European Symposium on Medieval Logic and Semantics, edited by Fabrizio Amerini and Irene Binini, Cahiers de l'Institut du Moyen-Âge Grec et Latin.
Parsons, Terence. 2008. The Development of Supposition Theory in the Later 12th Through 14th Centuries. In *Handbook on the History of Logic. Vol. 2: Mediaeval and Renaissance Logic*, edited by Dov M. Gabbay and John Woods, 157–280. Amsterdam: Elsevier.
Parsons, Terence. 2014. *Articulating Medieval Logic*. Oxford: Oxford University Press.
Parsons, Terence, and Graziana S. Ciola. 2025. The Traditional Square of Opposition. In *The Stanford Encyclopedia of Philosophy*, Summer 2025 Edition, edited by Edward N. Zalta and Uri Nodelman. https://plato.stanford.edu/archives/sum2025/entries/square/
Pasnau, Robert. 1997. *Theories of Cognition in the Later Middle Ages*. Cambridge: Cambridge University Press.
Pasnau, Robert. 2011. *Metaphysical Themes 1274–1671*. Oxford: Oxford University Press.
Pasnau, Robert. 2020. Medieval Modal Spaces. *Aristotelian Society Supplementary* 94 (1): 225–254.
Pasnau, Robert. 2025. Nominalism's Modal Paradise Lost. In *Oxford Studies in Medieval Philosophy*, edited by Peter King and Martin Pickavé, vol. 11. Oxford: Oxford University Press.
Quine, Willard V. 1948. On What There Is. *The Review of Metaphysics* 2 (5): 21–38.
Quine, Willard V. 1953. Three Grades of Modal Involvement. *Proceedings of the XIth International Congress of Philosophy* 14: 65–81.
Read, Stephen. 2015. Aristotle and Łukasiewicz on Existential Import. *Journal of the American Philosophical Association* 1 (3): 535–544.
Reina, Maria Elena. 2002. *Hoc hic et nunc. Buridano, Marsilio di Inghen e la conoscenza del singolare*. Florence: Leo S. Olschki.
Rodriguez-Pereyra, Gonzalo. 2019. Nominalism in Metaphysics. In *The Stanford Encyclopedia of Philosophy*, Summer 2019 Edition, edited by Edward N. Zalta. https://plato.stanford.edu/archives/sum2019/entries/nominalism-metaphysics/
Schabel, Christopher, and Charles Girard. Gregory of Rimini. 2022. In *the Stanford Encyclopedia of Philosophy*, Fall 2022 Edition, edited by Edward N. Zalta and Uri Nodelman. https://plato.stanford.edu/archives/fall2022/entries/gregory-rimini/

Spade, Paul Vincent. 1996. *Thoughts, Words and Things: An Introduction to Late Medieval Logic and Semantic Theory*. https://www.pvspade.com/Logic/docs/thoughts.pdf

Spade, Paul Vincent, Claude Panaccio, and Jenny Pelletier. 2025. William of Ockham. In *The Stanford Encyclopedia of Philosophy*, Summer 2025 Edition, edited by Edward N. Zalta and Uri Nodelman. https://plato.stanford.edu/archives/sum2025/entries/ockham/

Spruyt, Joke. 1994. Thirteenth-Century Discussions on Modal Terms. *Vivarium* 32 (2): 196–226.

Thijssen, Johannes M. M. H. 1999. The Commentary Tradition on Aristotle's *De generatione et Corruption*. An Introductory Survey. In *The Commentary Tradition on Aristotle's De generatione et corruptione: Ancient, Medieval, and Early Modern*, edited by J. M. M. H. Thijssen and H. A. G. Braakhuis, 9–21. Turnhout: Brepols.

Walz, Dorothea, and Reinhard Düchting, eds. 2008. *Marsilius von Inghen: Gedenkschrift 1499 zum einhundertsten Todestag des Gründungsrektors der Universität Heidelberg*. Heidelberg: Mattes.

White, Graham. 2018. Medieval Theories of Causation. In *The Stanford Encyclopedia of Philosophy*, Summer 2018 Edition, edited by Edward N. Zalta. https://plato.stanford.edu/archives/sum2018/entries/causation-medieval/

Wojtczak, Hanna. 2000. Jan Dorp—Autor komentarza do *Summulae logicales* Jana Burydana. In W kręgu filozofii klasycznej, edited by Bogdan Dembiński, Prace Naukowe Uniwersytetu Śląskiego w Katowicach, 224–243. Katowice: Wydawnictwo Uniwersytetu Śląskiego.

Zimmermann, Albert, ed. 1974. *Antiqui und moderni: Traditionsbewusstsein und Fortschrittsbewusstsein im späten Mittelalter*. Berlin: De Gruyter.

**Graziana S. Ciola** is Associate Professor of History of Philosophy at Radboud University Nijmegen. She specialises in medieval logic and semantics, with a particular focus on late medieval theories of consequence, modality, and impossibility. Her work investigates the interplay between logic, metaphysics, and epistemology in the later Middle Ages, and how these discussions contributed to the historical foundations of scientific and mathematical thought. She is the Principal Investigator of the ERC Starting Grant project The Impossible and the Imaginable: Late-Medieval Semantics of Impossibility and the Roots of Complex Mathematics (2023–2028).

# Chapter 6
# Pre-Copernican Astrology at Cracow, 1450–1500: A Still Unfolding Revolution in Our Understanding of Copernicus

**André Goddu**

**Abstract** This paper examines the teaching of astrology at Cracow in the second half of the fifteenth century as background to Copernicus's astrological activity. After a brief introduction, the paper summarises the teaching on spherical astronomy and the construction of horoscopes and astrological charts, considers Copernicus's astrological chart from around 1540, discusses Cracow astrologers in the second half of the fifteenth century, examines the discussion of the authenticity of Copernicus's annotations on Haly Abenragel's *In judiciis astrorum*, and then turns to the evidence for Copernicus's astrological activity and practice.

**Keywords** Astrology · Copernicus · Cracow University · Geocentric spherical astronomy · Horoscopes · Haly Abenragel · David Juste · Robert Westman

## 6.1 Introduction[1]

For over 100 years scholars rejected any notion that Nicolaus Copernicus practised astrology. In 1973 there were at least two suggestions, one by Richard Lemay and the other by Willy Hartner in an exchange with Edward Rosen, both at the 5th Centennial celebration of Copernicus's birth, that Copernicus would have studied and practised astrology, but these suggestions were almost universally rejected and

[1] This essay contains some corrections or qualifications of and additional material for an essay on the teaching of astronomy and astrology in a collection of papers from a conference in Frombork in 2023: Goddu (2025). That essay goes into more detail about Copernicus's handwriting, and the material on the teaching of astrology in particular was largely a summary of chapters by Mieczysław Markowski and Jerzy Dobrzycki (Dobrzycki 1975), some of the details of which I correct in this essay on advice from David Juste.

A. Goddu (✉)
Stonehill College, Easton, MA, USA
e-mail: agoddu@stonehill.edu



E. Jung et al. (eds.), *The Forgotten Giants: Overlooked Revolutions in Natural Philosophy*, Studies in History and Philosophy of Science 70,
https://doi.org/10.1007/978-3-032-34008-5_6

largely ignored.[2] A young Robert Westman attended that conference but he was not a direct witness to the exchange, and he initially maintained the standard view. A few years later, however, he announced a project that would challenge the orthodox view about Copernicus's exceptionalism in this regard. Lemay's account was based on the Cracow tradition of corrections to the *Alfonsine Tables*, which he believed also motivated Copernicus's involvement with astrology. Westman based his on Copernicus's relationship with Domenico Maria Novara, who was an active astrologer and issued many prognostications, and so it was plausible that Copernicus himself was trained and also involved in the astrological culture of the day. Later, Westman expanded his account to relate it to what he believed to be Copernicus's reaction to Adalbert of Brudzewo's linking of planetary qualities with their order in the geocentric system and to the publication of Pico della Mirandola's sceptical attack on astrology published in 1496.[3]

The thesis about exceptionalism might seem counterintuitive; after all what was more exceptional than Copernicus's adoption of heliocentrism? Historians, however, had extended Copernicus's exceptionalism in astronomy to exceptionalism regarding astrology and its culture, and that was indeed questionable. Copernicus had otherwise adopted the general contours of a cosmos created by a beneficent and providential God who provided mankind not only with revelation but with a universe endowed with qualities and motions that affect Earth. Westman's account, however, was not only contextual from the point of view of the history of ideas but also 'locally situated' in the power struggles of aristocratic and ecclesiastical courts.[4]

About fourteen years ago (2011), Westman fulfilled his promise with the publication of *The Copernican Question*.[5] Just to be clear, several reviews were positive,[6] but judging from the most cited reviews, I believe it fair to say that those critics who found it disappointing, flawed, and exaggerated tended to get the most attention. In effect, Westman had thrown a large rock into a pond that had ripple effects at first minor and almost anti-climactic but later of more impact and significance.[7]

Even as such critics lambasted his thesis, however, they simultaneously provided evidence of Copernicus's astrological activity, some of which Westman had overlooked, and, seemingly without irony, supported Westman's broader claim that Copernicus was not exceptional, that he was indeed trained in astrology, and attempted to contribute to its improvement. His critics' reactions went something like this: Of course, Copernicus studied astrology, and took it as a fact of ancient cosmography. The later ripple effects, however, have been more surprising. The new

[2] Ironically, Lemay's article (1978) appeared in the volume dedicated to Rosen.

[3] Albertus de Brudzewo (1900) (first published in 1494 and 1495) and Johannes Pico della Mirandula (1496). See Akopyan (2018, esp. 553, n. 36), where he cites *Disputationes* II, X, IV, 368–376; and Vanden Broecke (2003).

[4] See Cohen and Jacobs (2024). For a contrary perspective on political contexts, see Blumenthal (2014).

[5] Westman (2011).

[6] Exemplary are the reviews by Miguel Granada and Dario Tessicini. See Granada (2012) and Tessicini (2013). By contrast, see Swerdlow (2012, 2013).

[7] See Westman's account of the exchanges in 1973 in Westman (2026).

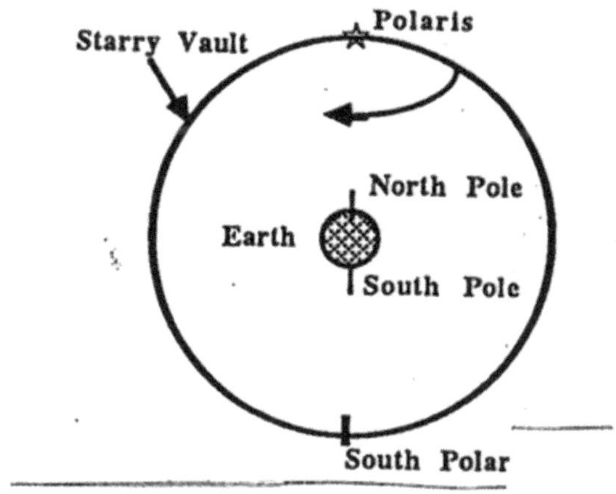


**Fig. 6.1** The celestial sphere. *Source* Crowe (1990, 1). All figures from Crowe used with permission of Dover Publications

picture, still incomplete, that is emerging is of a Copernicus who was an expert in astrology already as a student and continued to practice it until he died.

To express the fallout more sharply, most reactions were ambivalent: wrong about astrology as a motive for the heliocentric theory, right about Copernicus as not exceptional and, indeed, as an active astrologer.[8] The criticisms and rejections, however, stood out, thus stealing Westman's thunder, as we say, and depriving him unjustly of the credit he deserves for having resurrected the question about Copernicus's astrology, and for having stimulated and made legitimate a complete reconsideration of Copernicus's practice of astrology.

## 6.2 The Teaching of Geocentric Spherical Astronomy[9]

I begin with a description of geocentric spherical astronomy and horoscopes and astrological charts, that is, a brief account of their main features, restricting myself to basics.[10] Some technical terms are necessary, and I provide some illustrations to make these as clear as possible (Fig. 6.1).

[8] Of course, we can expect that some scholars with vestiges of Whiggish and even quasi-hagiographical inclinations will continue to support the views that both Birkenmajers, Ludwik and Aleksander, Grażyna Rosińska, and Edward Rosen defended, namely, of a Copernicus who was a harbinger of modern science, exceptional in his view of astrology, and unsullied by the antiquated cosmographical opinions of his contemporaries. More on these views later with citations. See Thorndike's pointed remarks concerning the lives of saints: Thorndike (1934–1941, 5:414). Cf. the edifying as well as entertaining account by Newman and Grafton (2001, 1–37).

[9] For background to this section, consult Eade (1984, 1–38), North (1986), Tester (1987, 238–240), Whitfield (2001), and Rutkin (2019). For a critical but sympathetic review of Tester, *History*, see Mary Ellen Bowden (1990).

[10] See Rutkin's "Excursus" (2019, lix–lxxxiv), which is partly a summary of *Speculum astronomiae* with some texts translated by Charles Burnett. Shorter and more digestible is Bezza (2014, 59–86).

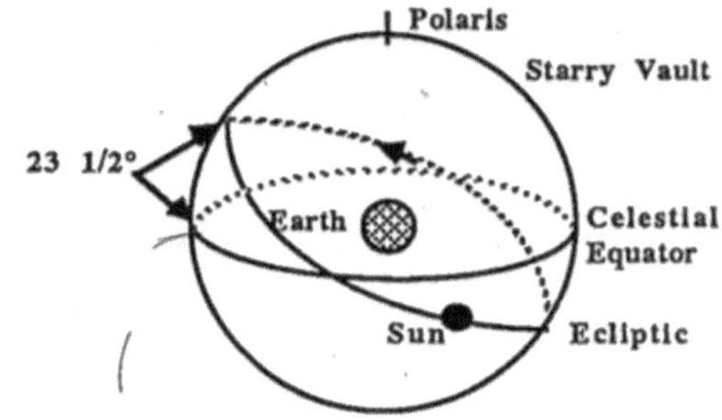


Fig. 6.2 The inclination of the equator to the ecliptic. *Source* Crowe (1990, 4)

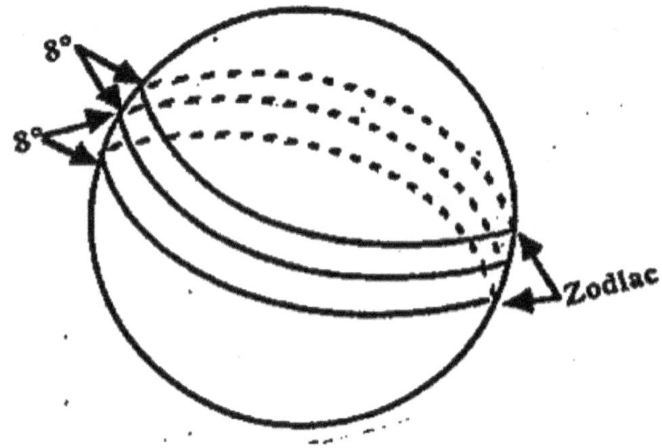


**Fig. 6.3** The zodiac. *Source* Crowe (1990, 11)

From a geocentric perspective we see the stars rise in the east and set in the west, and if we look northwards, we see the stars rotate over the course of the night in a counterclockwise direction. The daily rotation of the starry vault is transmitted to all the lower spheres, meaning that the entire cosmos rotates around Earth once a day[11] (Fig. 6.2).

The ecliptic is the circle on which the Sun is located on any given day of the year. That circle is inclined to Earth's equator at an angle of about 23.5°, and the inclination accounts for the differences in the seasons as the Sun crosses the equator and reaches the solstices, the tropics, that is, the stationary points where the Sun turns north or south (Fig. 6.3).

The zodiac is the region of constellations through which the Sun, planets, and Moon move. The ecliptic is the central circle of the zodiac. From northern latitudes, we see the Sun move in the southern sky west to east against the background of the

[11] For a clear explanation of some fascinating facts about geocentric spherical astronomy, see Tester (1987, 1–10). There is some slight variation in the number of hours, but all take approximately 24 hours for a complete rotation.

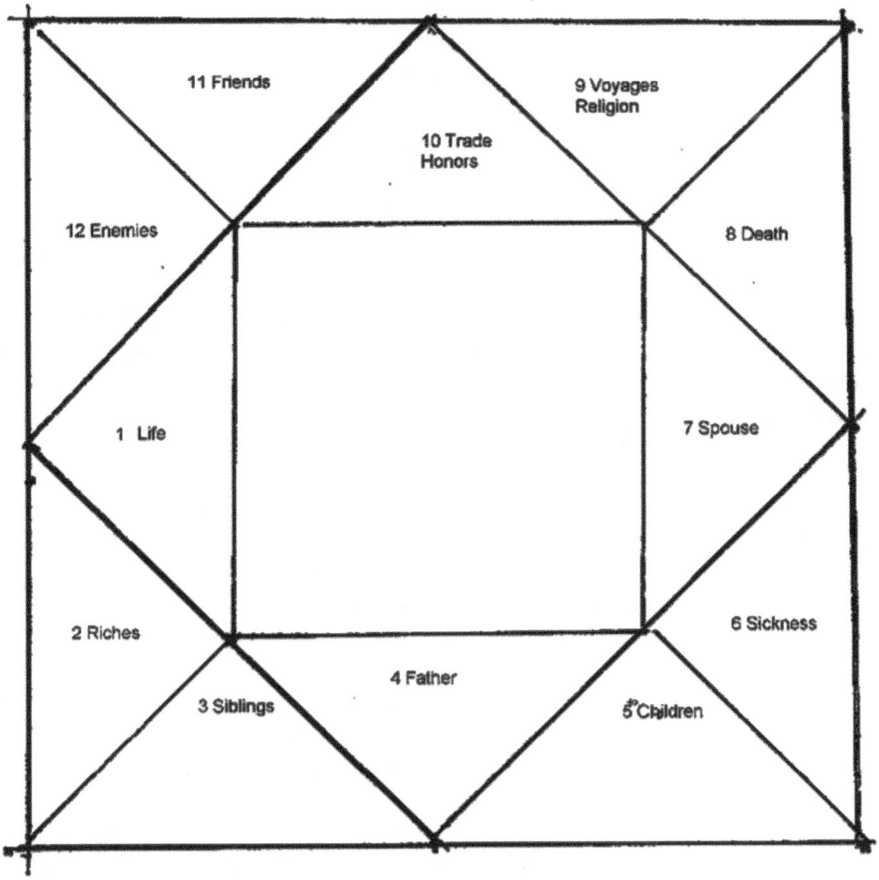


**Fig. 6.4** Astrological chart in square form

zodiacal constellations. The planets visible with the naked eye always appear to be within a range of about 8° north or south of the ecliptic, and the Moon within a range of about 5° north or south of the ecliptic. As the name suggests, eclipses occur when the Moon crosses the ecliptic. The northernmost point of the zodiac does not exceed 32°, the approximate northern latitude of North Africa. The significance of this detail is that European astronomers observed the planets, Sun, and Moon in the southern sky, which is why European horoscopes and charts are arranged with the South at the top, the East on the left, the West on the right, and the North at the bottom (Fig. 6.4).

The square chart is divided into quadrants, which are the cardinal points of a horoscope.[12] The term 'horoskopus' itself refers to the ascendant (1), the most important point, with the descendant (7), midheaven (10) culminating or zenith point overhead, and the lower midheaven (4) constituting the four cardinal points. The chart is further

[12] Whitfield (2001, 44).

**Table 6.1** Planetary symbols[15]

| | | | |
|---|---|---|---|
| ☉ Sun | ☽ Moon | ☿ Mercury | ♀ Venus |
| ♁ Earth | ♂ Mars | ♃ Jupiter | ♄ Saturn |

divided into twelve mundane or earthly houses placed in equilateral triangles, where a planet was 'lord' (*dominus*) of one or two signs and was said to be 'in exaltation' when in that sign.

One of the complications in describing the zodiac and horoscope is that the quadrants and houses cannot be assigned a fixed number of degrees, because of the inclination of the ecliptic to the equator and the latitude of the observer. The angles vary accordingly. In any particular chart, the locations of the constellations and planets in the houses depend on the date and time.

I will confine myself to only two more technical terms: aspects and the Lot of Fortune. An aspect is the angle between any two or more planets or points. Aspects can produce a number of geometrical figures—a square, triangle, or sextile, or they can be in conjunction or opposition. A triangle and a sextile were interpreted as harmonious because of the same gender, and a square as producing conflict because of opposite genders.[13]

The Lot of Fortune refers to a position as far from the ascendant as the angular distance between the Sun and Moon, and was considered fortunate because it combined the influences of the Sun and Moon. The occurrence of eclipses before or after the date of the chart also contained important information for interpretation of the chart.

Let us now consider one specific example, in which I will label the constellations among the twelve houses, and relate them to the planets. The traditional names of the planets were deities: Zeus (Jupiter), Kronos (Saturn), Ares (Mars), Aphrodite (Venus), and Hermes (Mercury). In addition, Apollo was the Sun. The powers or characteristics of the planets were associated with these deities and with their configurations and names[14] (Tables 6.1 and 6.2).

The example that I have selected is the horoscope drawn of Copernicus.[16] There were four types of horoscope: revolutions or general astrology, nativities (genethlialogy), interrogations, and elections. Until the sixteenth century the construction of horoscopes was normally kept distinct from their interpretation. Horoscopes of revolutions or general astrology refer to identification of all planetary conjunctions,

[13] Bezza (2014, 59–61), maintains that the origin of the aspects should be sought in Ptolemy's harmonic theory and musical intervals. Bezza cites Ptolemy, *Tetrabiblos* (1940, original ca. 150 C.E.), I, 9, but I. 13, where Ptolemy discusses the aspects of the signs and refers to harmonious and dissonant intervals, fractions, ratios, and proportions, seems more fitting.

[14] Whitfield (2001, 21–59).

[15] In the geocentric system, Sun and Moon are considered planets; in the heliocentric, of course, Earth is a planet with the Moon as its satellite.

[16] I have adapted and expanded the general description by Bezza (2014, 59–64).

**Table 6.2** Traditional constellation symbols

| | | | | | |
|---|---|---|---|---|---|
| ♈ Aries | ♉ Taurus | ♊ Gemini | ♋ Cancer | ♌ Leo | ♍ Virgo |
| ♎ Libra | ♏ Scorpius | ♐ Sagittarius | ♑ Capricorn | ♒ Aquarius | ♓ Pisces |

determination of annual revolutions, and the change of times and air (meteorology). Nativities relate to predictions of the course of life of an individual based on location and date and time of birth. Interrogations teach us how to judge a matter about which a question has been raised, especially whether it will come to pass or not. Elections deal with the choice of a propitious time to undertake an activity (journey, marriage, coronation, war, etc.)[17] In what follows I restrict my description to nativities in a basic and summary fashion.[18]

The apparent diurnal rotation of the stars, contrary to the counterclockwise rotation of the celestial sphere looking northwards, is clockwise from East to West. The annual rotation of Sun, Moon, and planets west to east appears counterclockwise.

The standard medieval and Renaissance horoscope represents the sky by a figure consisting of two squares: a central square and a large circumscribing square. In a nativity horoscope the central square contains the name of the person, date and time of birth, from which we can derive the latitude and place of birth. The large circumscribing square contains twelve equilateral triangles that divide the apparent diurnal motion of celestial bodies around Earth into the twelve mundane or earthly houses. Within a given period, each house has its own particular force and quality. The first four houses beginning along the horizon and meridian lines with the bases facing the central square are located clockwise beginning with the first house in the East, and are numbered 1, 10, 7, and 4. These four houses are followed by the houses in triangles whose bases face the higher and lower sides of the external square, and they are numbered 11, 9, 5, and 3. These houses have a lesser degree of force. The final four houses with the bases of their triangles facing the eastern and western sides of the external square are the weakest, and they are numbered 12, 8, 6, and 2 with 12 and 2 on the eastern side and 8 and 6 on the western side.

The locations of the planets were derived from ephemerides, tables of planetary positions on specific days for a given year or over a given period.

Copernicus's chart appears to have been constructed for Toruń at about 53° North latitude, for the equator has turned to bring the constellation Aries above the horizon[19] (Fig. 6.5).

[17] Compare Rutkin (2019, lix–lxxxviii), Bezza (2014, 60–71), Whitfield (2001, 29–33 and 55–66), and North (1986, 1–51).

[18] See Rutkin (2019, lxx, note 31). The observation concerning the distinction between construction and interpretation is based on David Juste's extensive familiarity with the sources.

[19] Cf. Tester (1987, 75, Figure 10b), a spherical representation with north at the top.

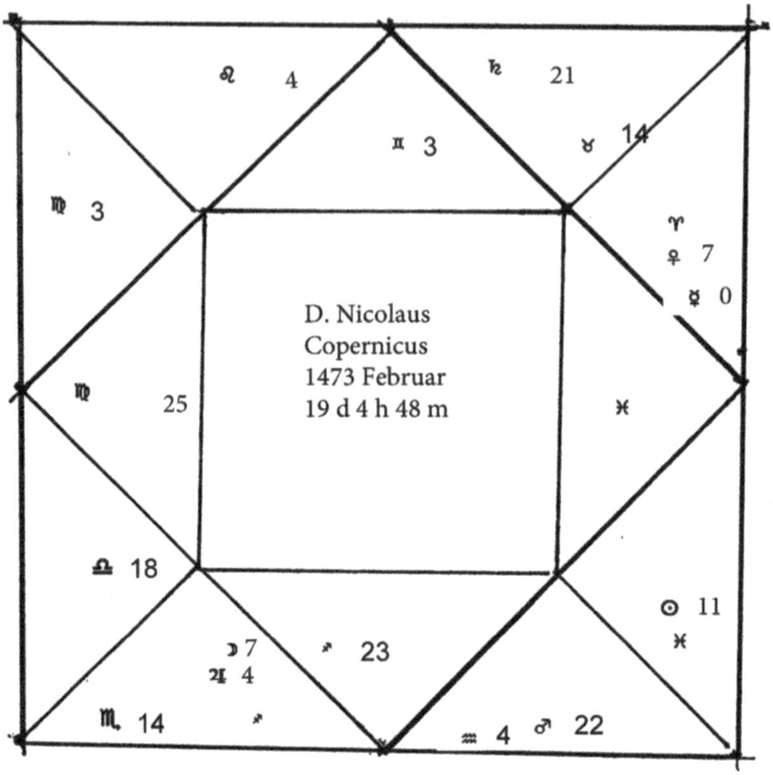


**Fig. 6.5** Copernicus's chart[20]

## 6.3 Interpreting Astrological Charts[21]

Ptolemy believed that experts in astronomy and skilful practitioners in the art could produce some reliable information useful for, among others, farmers, navigators, physicians, and counsellors to political leaders. As with astronomical models, he believed that with time some progress in prognostication was possible.[22] One can hardly doubt his sincerity, yet a startling statement about universal astrology in Book II, 8 expresses caution about the difficulties involved and surely posed a challenge to those undertaking such a daunting task:

[20] My transcription of Codex lat. Monacensis No. 27003, folio 33$^{v}$. A photographic reproduction is in Biskup (1973, Illustration 22, after p. 192).

[21] By the time, around 1540, that Copernicus's chart was drawn up, the author probably had access to the Greek edition of *Tetrabiblos* edited by Johannes Camerarius in 1533. See Ptolemy (1940 (Robbins edition), p. XIII and II. 8, p. 189). Compare with Swerdlow and Neugebauer (1984, I, 454–457) and Westman (2011, 115–116, 535 n. 50).

[22] Ptolemy (1940 (Robbins ed.), x–xii and I. 2–3). Ptolemy's *Syntaxis*, better known in the West as *Almagest*, was also addressed to Syrus.

> It is of course a hopeless and impossible task to mention the proper outcome of every combination and to enumerate absolutely all the aspects of whatever kind, since we can conceive of such a variety of them. Consequently, questions of this kind would reasonably be left to the enterprise and ingenuity of the mathematician, in order to make the particular distinctions.[23]

In another text dealing with medical prognostication, Ptolemy distinguishes between the certainty of general astrology and the relative uncertainty of prognostications of single events in part because earthly events depend on conditions that are more proximate than the stars.[24]

Based on David Juste's likely conjecture that Copernicus himself constructed the chart, I had originally considered interpreting it according to Ptolemy's guidelines,[25] allowing first for Ptolemy's own cautions and qualifications:

> [...] among all genethlialogical inquiries whatsoever, a more general destiny takes precedence of all particular considerations, namely, that of country of birth, to which the major details of a geniture are naturally subordinate, such as the topics of the form of the body, the character of the soul and the variations of manners and customs, it is also necessary that he who makes his inquiry naturally should always hold first to the primary and more authoritative cause, [...].[26]

Further complications, however, led to the realisation that any worthwhile interpretation would require a separate and lengthy paper. Accordingly, here I mention some of the requirements for a professional interpretation, preceded by some additional facts regarding Copernicus's chart.

On the assumption that Copernicus authored it, Juste supposes that Rheticus took a copy of the chart with him back to Wittenberg, and Erasmus Reinhold made the copy of it that we saw above and included it in his collection of horoscopes. Later readers of Ptolemy distinguished at least seven different methods of calculation, but we may assume that the astrologer followed the method advocated by Regiomontanus.[27] That method known as the equatorial method with moving boundaries divided the intervals between the angles of the signs by means of double temporal hours, counting either from midnight to noon or noon to midnight, making its one-third division along the

---

[23] Ptolemy (1940 (Robbins ed.), 189). In a footnote, Robbins interprets 'mathematician' (μαθημαικος) as 'astrologer.'

[24] Ptolemy (1940 (Robbins ed.), 30–31), a text that Robbins characterises as highly condensed and obscure. He cites the *Paraphrase* by Proclus to support his translation. Compare Lerch (2015, 34–35, notes 44–47), who cites Robbins and Girolamo Cardano's commentary, which is explicit about the proximity of earthly causes. Cardano expresses great caution about prognostication of particular events.

[25] David Juste informed me of his suspicion in a private email message. In my conference presentation I offered two possible interpretations, the first based on Birkenmajer's conjecture that Johannes Schöner might have constructed the chart possibly for the purpose of establishing that Copernicus was foolish with a tendency to madness, and a second based on Juste's conjecture, relying on Ptolemy. See Birkenmajer (1900, 410–411). Both readings could be based on *Tetrabiblos* III. 13.

[26] Ptolemy (1940 (Robbins ed.), IV. 10., 439).

[27] On the seven methods, see Bezza (2014, 64), and North (1986, 46–47). Cf. Eade (1984, 42–47), where he describes and compares six systems or methods, and 225 for the mathematical formula for the equatorial method.

ninety degrees of the celestial equator that lie between the tenth house and the first house.[28] In Copernicus's chart, the time was counted from noon, yielding 4:48 p.m. There are some problems with its accuracy. Swerdlow and Neugebauer have shown that the time does not agree with some of the parameters but that they are consistent with a time about two hours later. They suggest that Copernicus's selection of the time was schematic, that is, based on a time that is exactly one-fifth of twenty-four hours.[29]

Beginning with a general orientation, we consider Copernicus's temperament to fall under the northern European type, and the signs arranged such that the ascendant was in Virgo and House 1, Saturn and Gemini in upper mid-heaven and House 9, Venus and Mercury in Aries in House 8, the Sun in Pisces in House 7, Mars in Aquarius in House 5, the Moon, Jupiter, and Sagittarius in lower mid-heaven in House 3. Notice, however, that some constellations occupy more than one house, and the chart omits Cancer and Capricorn, the ascending and descending lunar nodes, and the Lot of Fortune (the *Pars fortunae*).

A complete interpretation requires the following additional variables: the planetary aspects, prediction of syzygies and the most recent eclipses, the entry of the Sun into the four cardinal points (especially the vernal equinox), the relation of the terrestrial houses to the zodiacal signs, male and female signs, and the Lot of Fortune among others. Indeed, because of the need to adjust a given table to the latitude of Toruń, the possibilities of permutation, the effect of the seven planets in any of the twelve signs and in any of the twelve houses amount to 1008, each with its fortunate or unfortunate potential.[30]

We may be astonished to realise that despite all the complications and difficulties that astrologers encountered, they continued to believe in the possibility of some foreknowledge of—and the possibility of their subjects' partial control over—their destinies.

---

[28] Eade (1984, 45).

[29] Swerdlow and Neugebauer (1984, I, 454–457), do not offer any further explanation of why Copernicus would have chosen one-fifth. They point out, however, that the midheaven Gemini 23° [sic] and the ascendant Virgo 25° with the Sun at Pisces 11° agree with a time of about 6:40 pm. It is not clear whether 23° is a typographical error or a correction. See also the photographic reproduction in Wasiutyński (1938, after p. 496), which clearly reads Gemini 3°. No horoscope corrects the time even though the given parameters indicate a time after sunset and around 6:40 pm. Mikulski (2023, 204–205), suspicious of the accuracy to the minute wonders if even the date is reliable. Birkenmajer (1900, 407–410), cites another horoscope that puts the day at 10 February, probably a copyist's mistake, but otherwise the year is identical and time, 4:38, nearly the same. Note also that Mikulski mistakenly sets the time at "4:48 AM," but he, like David Juste, also thinks Copernicus constructed his own horoscope.

[30] See Eade (1984, 39). Perhaps because of such considerations, no competent historian of astrology, to my knowledge, has even attempted an interpretation. With the help of an expert I may in future select two 'routines' and attempt a precise interpretation. All historians remark on the difficulties. For example, *see* Tester (1987, 174), just regarding the terrestrial houses. On the frustration over inaccuracy, see Kremer (2006).

## 6.4 Pre-Copernican Astrologers at Cracow

Around 1405 a chair in mathematics and astronomy known as the Stobner chair was established at Cracow, just a few years after the re-founding of the university about 1397–1401.[31] Stobner professors were charged with interpreting the *Alfonsine Tables*, and preparing almanacs and ephemerides, which suggests that duties and aims were also astrological.[32] Prior to 1405 the sources are fragmentary, such that it is often difficult to classify their activities.

In 1459 a chair specifically of astrology was founded from a bequest by Marcin Król of Żurawica. The subject was taught in a manner popular in the Middle Ages on the basis of Ptolemy's *Tetrabiblos* (*Quadripartitum* in a Latin translation of an Arabic translation),[33] through commentaries and exercises on the *Alfonsine Tables*, and various works of Regiomontanus on calendars and tables. Jerzy Dobrzycki concluded that the culmination of lectures on astronomy at Cracow in the late fifteenth century was the course on astrology, based largely on Ptolemy's *Quadripartitum.*[34]

Although Cracow astrologers were trained to make observations and construct tables or correct available tables, they relied for the most part on pre-calculated tables, at first in manuscript and from around the 1480s available in print. Because of reliance on pre-calculated tables, many astrologers used tables constructed for other latitudes that were inappropriate for the construction of a horoscope, and, hence, were worthless.[35] Teachers' principal tasks were commentaries on and the application of Ptolemy's rules in the *Tetrabiblos* and the Canons of the *Alfonsine Tables*, along with the construction of tables prepared for the Cracow meridian (*Tabulae resolutae*), the use and adaptation of John Bianchini's latitude tables, the construction of horoscopes, for which they also relied on Ptolemy's physical explanation of the transmission of stellar and planetary powers to terrestrial bodies affecting weather, climates, and the four humors of Greek medical theory. Note, however, that Cracow astronomers were quite cautious about adapting tables to their local meridian and latitude.

Historians of astrology refer to the activity of mathematical or theoretical astrologers as scientific, by which they refer mostly to the use of the *Almagest* to construct tables of planetary positions and mathematical models to predict future positions.[36] Accordingly, they were trained in geometry and trigonometry, both plane and spherical, to accomplish their tasks, engaged, in other words, in what other historians have called 'exact science.' In addition, they were trained in natural philosophy to identify the characteristics of zodiacal and planetary signs, so that in combination with their calculations and application of pre-calculated tables they were trained to construct horoscopes.

---

[31] See, for example, Moraw (2008, 181–206) and Lemay (1978).

[32] See Rosińska (2006, 9–28).

[33] Ptolemy (1940 (Robbins ed.)).

[34] Dobrzycki (1975, 1:128): "Ukoronowaniem wykładów astronomii był kurs astrologii, oparty na *Czteroksięgu* Ptolemeusza na traktacie astrologicznym Haly Abenragela."

[35] See especially Whitfield (2001) and North (1986).

[36] Here again points made repeatedly by Whitfield and North, as well as Rosińska (2002).

The applications of mathematical calculations and their interpretations fell under the artful dimensions of the astrologers' craft, entailing the exercise of individual judgment where errors were more subjective and more difficult to correct. The reasons for these differences varied; astrologers employed diverse rules and used diverse tables that led to different results.[37]

The remainder of this account is tentative, selecting examples based primarily on Polish secondary literature, and suggesting further study of archival materials. The proposals offered, however, are also suggestive of possible influence on Copernicus. Robert Westman's hypothesis about a link between Adalbert of Brudzewo's description of planetary qualities and the ordering of the planets requires further textual support.[38] Brudzewo did not explain how the stellar and planetary qualities are transmitted to other planets and Earth in his *Commentariolum* (1495), a topic that he presumably treated in his still unedited commentary on Ptolemy's *Tetrabiblos*.[39]

With the mid-fifteenth-century reform of the university curriculum, Marcin Król of Żurawice and the next generation of astronomers at Cracow (1450–1475) established the reputation of the university as one of the leading centres of astronomy and mathematics in Europe down to the early sixteenth century. Some of the masters went on to obtain degrees in medicine or even theology in some instances and lecture at other universities; nevertheless, with their teaching on mathematics, astronomical tables, eclipses, and their commentaries on Ptolemy's *Quadripartitum*, they re-introduced Cracow scholars to the Greek tradition, supplemented by their introduction of the tables and works of Peuerbach, Regiomontanus (tables and *Calendarium*), and John Bianchini (latitude tables). There were lectures and treatises on the *Alfonsine Tables*, some of them correcting the tables or adjusting them to Cracow longitude and latitude, the calendar (*computus*), the theory of proportions (Boethius), the calculation of fractions (*Algorismus minutiarum*), geometry (Euclid), changes in the weather (*De mutatione aeris*), *Electiones astrologice*, and other texts that prepared students for the practice of astrology in relation to meteorology and medicine.[40]

The peak of the development of what Markowski calls the 'Cracow astronomical school' occurred in the last decade of the fifteenth century when two outstanding Cracow masters, John of Glogovia and Adalbert of Brudzewo, were particularly active.[41]

John was not an expert in astronomical mathematics, but wrote primarily important and popular introductions to elementary astronomy. He was the author of *Computus chirometralis*, which became quite popular in the university environment at the

[37] Once again, a point made abundantly and frequently by Whitfield and North.

[38] Westman (2011) and Westman (2013, 2017).

[39] We will return to these questions later under Copernicus's astrological activity and practice.

[40] Markowski (1975a). See the extensive lists of works cited by Rosińska (1984 of Iohannes de Glogovia, 531, and of Albertus de Brudzewo, 523), most of which remain unedited. On John, see also Thorndike (1934–1941, 4:449–453; 5:160, 170, 179–180, and 333). On Latin manuscripts of works on Ptolemy in Cracow, see also Juste (2016).

[41] Markowski (1975a).

beginning of the sixteenth century. Already before 1483, John wrote a commentary on the medieval work on theoretical astronomy, the *Theorica planetarum* by Gerard of Sabbionetta.[42] Around 1488 he prepared the *Interpretatio 'Tabularum resolutarum ad meridianum Cracoviensem.'* In 1492 John composed *Tractatus de 48 imaginibus caelestibus*, providing a description of the constellations and their astrological significance.

Although not an astronomer or mathematician, he could use tables, and made some observations of lunar and solar eclipses as well as the conjunction of Saturn and Jupiter in 1504. In astrology, however, he wrote another important treatise, *Persuasio brevis quomodo astrologiae studium religioni Christianae non est adversum.*[43] As the title indicates, it is an apologia for the study of astrology as not contrary to the Christian religion.

John wrote an extensive *Astrological Summa*, which consists of four parts: an introduction to astronomy (*Introductorium in artem astronomiae*), a treatise on the change of weather (*Tractatus praeclarissimus in iudiciis astrorum de mutatione aeris*), the third on horoscopes (*Introductorium in scientiam nativitatum*), and the fourth on the selection of propitious days (*Tractatus de electionibus*). Each part constitutes a self-contained whole. The first part is known only from a number of manuscript copies. The second part was published in Cracow in 1514. The third part on horoscopes and the fourth part of the astrological compendium on judicial and *catharchic* astrology have not yet been published.

The approach of Saturn and Jupiter in 1504 became an opportunity for John to write a special treatise *Utrum stellarum et planetarum figurationes caelestes variae anno Salvatoris 1503 et 4 venturae rerum mirabilium adventum.* John is also one of the astrologers who prepared the *iudicium* about the comet in 1472. Such prognostications in Latin and German were printed not only in Cracow but also in Leipzig and Merseburg. Prepared annually, they were of interest normally for only a year and so not always preserved in subsequent years. Of these only a small number have survived.

Adalbert of Brudzewo was more active in astrology than earlier historians have acknowledged. In 1482 Adalbert edited *Tabulae resolutae pro supputandis motibus corporum caelestium.* This work, which is an explanation of the astronomical tables reduced to the Cracow meridian, enjoyed great popularity at the end of the fifteenth and sixteenth centuries at the University of Cracow. Earlier in 1478 Adalbert copied the Canons to the Tables of Directions and Movements of Heavenly Bodies (*Tractatus de directionum et profectionum*) by Regiomontanus.[44] On the basis of these tables, which he slightly expanded in several places, Adalbert also prepared the *Tabulae duodecim domorum caeli*, which is composed of the tables and the text explaining

[42] Pedersen (2011, 18) with reference to Pedersen (1962, 225–232), attributes the text to an anonymous author.

[43] I owe this reference to Juste (2023/2027). See Jan z Głogowa (1985) and Markowski (1975b, 107–110).

[44] Copernicus's *Uppsala Notebook* (Uppsala, Copernicana 4) included an edition of Regiomontanus's *Tabulae* published in 1490 as well as an edition of the *Alfonsine Tables* published in 1492.

the terrestrial houses, one of the fundamental exercises for relating the earthly houses to the zodiac. For the year 1481 he calculated the Astronomical and Astrological Calendar. Only a fragment has survived from his Calendar for 1491 or 1492.[45] Among additional works of Brudzewo are the *Summarius parvularum computalium* and the table giving the location of the fixed stars, *Loca ortusque stellarum fixarum*. Adalbert also wrote a meteorological work, which was used by James of Iłża in 1519, and a treatise entitled *Conciliator*, owned in 1507 by Martin of Cracow.

Engaged primarily in theoretical investigations, Adalbert demonstrated that he was proficient in astrology through the texts cited above, a note on the plague from 1482, an extensive forecast for 1487, and a horoscope written in 1481 for John Morzystyn's newborn daughter.

## 6.5 Copernicus's Annotations in Haly Abenragel's *In judiciis astrorum* in Uppsala, Copernicana 6[46]

In a major study of Copernicus's library published in 1978, Paweł Czartoryski concluded that annotations in an astrological text entitled *In judciis astrorum* by Haly Abenragel were inauthentic.[47] Czartoryski declared: "The text of [...] Haly's astrological work, contrary to previous identifications, was annotated by a different hand which we shall call 'hand B' [...]."[48] Unfortunately, Czartoryski did not state the criteria by means of which he reached this conclusion, nor has 'hand B' ever been identified. Nor did he explain the curious fact that 'hand B' appears, so far as we know, only in books owned or used by Copernicus.[49] In fact, with one exception cited below, Stefan Hartmann, most judgments about Copernicus's hand are impressionistic and anecdotal without a description of the characteristics that justify the conclusions.

I had earlier conjectured that the annotations that Czartoryski identified as 'hand B' were, in fact, likely written by Copernicus in the 1490s during his Cracow period

---

[45] Markowski (1975b, 111–115).

[46] This section is supplemental to the paper cited in footnote 1, containing corrections and additions.

[47] Czartoryski (1978, 365–366). The paper was originally delivered at the 1973 conference commemorating the 500th anniversary of Copernicus's birth. See also Goddu (2004).

[48] Czartoryski (1978, 366). The identification of the author of *In judciis astrorum* contains a confusion. According to David Juste in a private email message, the full name of the author was Abū al-Ḥasan ʿAlī ibn Abi al-Rijāl, where 'Abū al-Ḥasan' = 'Albohazen' and 'ʿAlī ibn Abi al-Rijāl' = 'Haly Abenragel' in Latin. In other words, the author is sometimes called Haly Albohazen or Haly Abenragel in Latin. Thorndike (1934–1941, 5:419), had already pointed out, citing Leopold Prowe, the Albohazen Haly text and Copernicus's annotations from the *Quadripartitum*. Cf. Prowe (1883–1884 (repr. 1967), I, ii, 414; II, 256–258). As Juste points out, however, all transcriptions prior to his own are partial and defective.

[49] Of course, this fact could be due to coincidence and the small sample; the discovery of a book neither owned by nor associated with Copernicus and containing annotations in 'hand B' would require some re-evaluation.

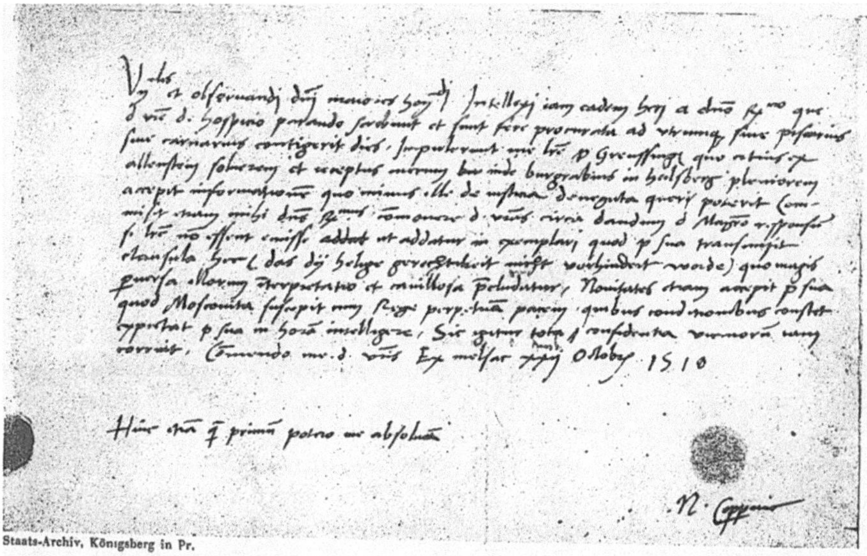


**Plate 6.1** Photograph of letter by Copernicus to a fellow canon of Mehlsack, 22 October 1518. *Source* Wasiutyński (1938, after p. 280)

or in Italy prior to 1504.[50] Rosińska evidently arrived at the same conclusion,[51] and David Juste has provided further evidence that leads me to advance this conclusion as the most likely explanation for Czartoryski's mistake, namely that 'hand B' is Copernicus's early Cracow hand, and that as he began changing his handwriting style in Italy, he reverted to his Cracow hand from time as late as 1518. Copernicus's annotations are difficult to date because we can conclude at best that he wrote them after he acquired a book. Correspondence, however, is datable, and one example of a letter from 1518 indicates that he had by that time adopted a double-l ligature that he likely used earlier and that later became standard in many annotations, but in that same letter he also separated the two letters clearly (Plate 6.1).

In the first line, note the separated letters 'll' in what appears to be the word 'Intellexi,' and in the second to last line, the double-l ligature in the word 'intelligere.'[52] This example is consistent with other cases where he reverted to earlier

[50] See Hartmann (1973), who demonstrated changes in Copernicus's handwriting from dated correspondence; and Goddu (2025). Compare Wasiutyński (1963, 67–71), Rosińska (2002, 119–121), and Juste (2023/2027). I am grateful to Juste for his having sent me a version of his paper presented at a conference in Toruń. He is preparing a version that is scheduled to appear in 2027 in *Studia Copernicana* 47, forthcoming.

[51] See Rosińska (2002, 120).

[52] Hartmann did not record the double-l ligature among his examples.

forms of writing after he adopted a newer style.[53] The point is that the separation of the letters 'll' is no proof that the annotations in the Haly text were inauthentic.

## 6.6 The Evidence Almost Nobody Examined

Not nearly as dramatic or as abundant as in *The Book Nobody Read*,[54] the evidence for Copernicus's astrological activity and practice has been available for quite some time.

Information about the teaching of astrology in the Cracow curriculum has been available for decades, yet it was largely dismissed as having no relevance for Copernicus's training. It is possible that Copernicus received instruction from John of Glogovia and Adalbert of Brudzewo, although it is likelier that he attended the lectures of their colleagues and students. In particular, we can point to Albert of Szamotuli's lectures on Ptolemy's *Quadripartitum* and on astrology, Bernard of Biskupie's lectures on the *Table of Eclipses*, and the lectures of Albert of Pnyewy, Simon of Sierpc, and Michael Falkener of Wrocław. Markowski counts all the authors mentioned here as students of John of Glogovia and Adalbert of Brudzewo.[55]

Recall that Copernicus shared the ancient perspective of a cosmos that is finite, unified, and whole, in which the whole determines the existence and places of the parts, all influencing one another.[56] As we noted earlier, the acceptance of a cosmos inhabited by living beings consisting of four humours entailed notions of stellar influence and physical transmission, ideas that we have reason to believe Copernicus also shared.[57]

The evidence for Copernicus's activity in astrology consists principally of four items. First, his comments on Haly Abenragel's *In judiciis astrorum* show, according to David Juste and contrary to Ludwik Birkenmajer, that he was proficient in astrology. Second is evidence of arranged observations with colleagues in Cracow of celestial events that were conducted mainly for astrological purposes. Third, a letter from his friend Bernard Wapowski, attesting to the fact that Copernicus had prepared ephemerides for astrological purposes, and another from Johannes Apelt, referring apparently to Copernicus as a well-known astrologer, support conclusions about his

[53] Compare Biskup (1973, 88, item 167), with Wasiutyński (1938, 280) and Hartmann (1973), tables, added as appendix in Goddu (2025).

[54] Gingerich (2005), his famous account of the discovery of hundreds of copies, several with annotations.

[55] In Goddu (2010, 159–161), I proposed a list of lectures at the University of Cracow in the period 1491–1495 that Copernicus could have attended, all based on John of Glogovia's recommendation for the order in which the lectures should be given and on the list of classes in the documents: *Liber diligentiarum* (1886, 22–28) and Karliński (1873, Appendix III, Tables 2–8). See also my references to Ludwik Birkenmajer, Aleksander Birkenmajer, Jerzy Dobrzycki, Marian Biskup, Grażyna Rosińska, and Markowski (1975b, 115–125).

[56] Goddu (2009).

[57] See Keyser (2018).

astrological expertise. Fourth, following on a suggestion by Robert Westman, the very title *Revolutions* suggests that Copernicus intended his reform of Ptolemaic astronomy to contribute to the reform of astrology.

In his thorough analysis of Copernicus's annotations in the Haly text in the translation from the Arabic by Egidius de Tebaldis, in which he fully credited Robert Westman for having revived the discussion of Copernicus's astrological practice, David Juste originally believed that Copernicus's comments on Haly relied on the translation of Ptolemy's *Tetrabiblos* by Plato of Tivoli. Juste has since expressed doubts about Copernicus's use of this translation. Furthermore, this translation has never been mentioned among those in Copernicus's library. Although this and other details remain obscure, we may glean the following from Juste's analysis. Juste focuses first on the annotations that show Copernicus interpreting birth charts, expanding on both Haly and Ptolemy, and demonstrating his mastery of not only their texts but also presumably having one or more birth charts in front of him as he made his additions. Copernicus used the astrological symbols for the planets and other details that are not found in Haly and Ptolemy but rather typically in tables, almanacs, and ephemerides, which again shows that he was familiar with the technical details and able to read the charts critically.

As for Copernicus's additional sources, it seems likely, based on the testimony of Copernicus's early Polish biographers, that Copernicus owed his accomplishments to the University of Cracow, which suggests that Copernicus became an expert under the influence of Cracow masters and other students.[58]

Cracow professors used the *Quadripartitum* translated by Egidius of Tebaldi either in manuscript or the 1493 printed edition exclusively. Cracow commentators who used the Egidius translation may have instructed Copernicus. Contrary to Juste's initial view, I think it unlikely that he had no familiarity with the Egidius translation while he was still in Cracow. Juste himself notes that Copernicus relied on other unknown sources for some of his comments.[59]

Second, on the basis of comments in *De revolutionibus*, we know that Copernicus reduced all calculations of observations to the Cracow meridian:

> All these values [are reduced] to the meridian of Cracow. For Gynopolis, which is commonly called Frombork, where we generally made our observations, is located on the mouths of the Vistula River and lies on the meridian of Cracow, as we learn from lunar and solar eclipses observed simultaneously in both places.[60]

Evidence supplied by Ludwik Birkenmajer indicates that Copernicus arranged simultaneous observations of eclipses in both cities.[61] Because of the relation of the observations to the theory of the inner planets and the Moon, Birkenmajer, thinking that

[58] See Hilfstein (1980).

[59] Juste (2023/2027). Again in another private communication, Juste has since discovered reasons to think that Copernicus did indeed know the 1493 edition perhaps the copy owned by Domenico Maria Novara. In a later message, he asserted that he found no evidence in Novara's copy that Copernicus had seen it.

[60] Copernicus (2015, IV. 7) and Copernicus n (1978, 191) [translation slightly modified].

[61] Birkenmajer (1900, chapter 20, 449).

the final redaction of the last three books was well underway, listed several solar and lunar eclipses made before 1525. Of these, it is not possible to say with certainty which were pre-arranged, but some are likelier than others, especially those that can be associated with Copernicus's friends and colleagues.

Although Copernicus refers to some observations in *Revolutions* and in his correspondence, nobody has examined the sources in detail to try to determine which might have been pre-arranged. Citing Simon Starowolski's *Hekatontas* and information from Martin Radyminski, Birkenmajer suggested very cautiously that Copernicus might have arranged observations with Bernard Wapowski, Nicolaus Szadek, and Martin of Ilkusz.[62] Furthermore, Birkenmajer cited a Cracow incunabulum with extensive annotations by Martin of Ilkusz, whom Birkenmajer identifies with Martin Bylica of Olkusz, also known as Martin Biem of Olkusz. Among the annotations are lists of cities with their longitudes and latitudes, including conspicuously those for Cracow, some of which appear to be based on observations made before 1531.[63]

Birkenmajer knew two manuscripts of *Analecta astrologica*, one of which contained a horoscope by Bernard Wapowski, and the other containing comments by Martin Biem enumerating Wapowski as his source for some astrological details.[64] Birkenmajer cited observations of Martin Biem's solar and lunar observations, including a reference to a "conjunction of Saturn and Jupiter according to Copernicus."[65] Birkenmajer completes this chapter with citations of more Cracow observations, concluding with some cautious conjectures relating to Copernicus. In Chapter 26, Birkenmajer examined Copernicus's copy of Johannes Stoeffler's *Calendarium romanum magnum*, in which Copernicus recorded six solar and six lunar eclipses, noting Copernicus's apparent location, and recording observations made by Cracow colleagues, Martin Biem and Nicolaus Szadek.[66]

Birkenmajer, of course, believed these to be of exclusively astronomical or geographical significance for Copernicus, even though he knew that Copernicus's correspondents were definitely motivated by astrological concerns and even knew of their astrological annotations. Their purpose, as Birkenmajer asserted, was to ascertain the longitude of Frombork and Cracow, but that determination, as we saw, served to construct astrological charts.[67]

Third, David Juste's comments on the letters by Bernard Wapowski and Johannes Apelt suffice to add confirmation of Copernicus's astrological activity and about the relevance of his theory for astrology.[68] In his letter from October 1535, Wapowski

---

[62] Birkenmajer (1900, compare 317–319 with 449–453).

[63] Birkenmajer (1900, 453–455).

[64] Birkenmajer (1990, 455–459).

[65] Birkenmajer (1990, 459–462).

[66] As Birkenmajer (1990) explained at the end of Chapter 20, he completed that chapter before he discovered Copernicus's copy of Stoeffler's *Calendarium*.

[67] For authors including Birkenmajer who dismissed Copernicus's early astrology as youthful and purely scientific, see Rosińska (2002, 121). For his part Rosen dismissed the Haly annotations as inauthemtic based on Czartoryski's erroneous judgment. See Rosen (1984, 111–112).

[68] Juste (2023/2027).

commends an almanac that Copernicus prepared on the basis of new tables derived from his theory, offering corrected motions of the planets.[69] That the almanac related to astrology is confirmed by Wapowski's caution that Copernicus did not have time to complete the 'aspects,' a feature that is of purely astrological significance. Wapowski also stresses the importance of the almanac for the astrological community.

Note that Wapowski confirmed Copernicus's practice of astrology. It is extremely unlikely that he would have misrepresented the views of his friend, whether deliberately or inadvertently.

A few months earlier, Johannes Apelt addressed a letter to Duke Albrecht of Brandenburg that included four horoscopes without an interpretation, and he added that if the duke could not find a suitable astrologer, he should consult 'an old canon of Frauenburg,' whom Andreas Kühne and Juste identify with Copernicus.[70] That interpretation is plausible on the grounds that Albrecht later consulted Copernicus for medical advice. Again, Juste concludes:

> The implications of Johannes Apelt's letter, if we trust his words, are that Copernicus was still reading horoscopes forty years after his studies at Cracow and that this astrological activity was known outside Frombork.[71]

Juste, then, summarises his case:

> It remains of course possible that Copernicus was somehow misrepresented or abused by Bernard Wapowski and Johannes Apelt, but when we consider these two letters, together with Copernicus's annotations to Haly, a new picture emerges. Copernicus was already skilled in astrology when he was a student at Cracow and evidence suggests that he never ceased to cast and interpret horoscopes. However surprising this may appear, it is, in fact, no more than a banal trajectory for a European astronomer of the fifteenth and sixteenth century.[72]

Of course, we cannot be certain that Apelt was referring to Copernicus, nor do we have any evidence that the duke followed his advice. Perhaps the duke did know a suitable interpreter. Other than Copernicus we know of no 'old canon of Frauenburg' who was a suitable astrologer or astronomer for that matter.

Fourth, the title of Copernicus's masterpiece, *De revolutionibus*, seems to offer further evidence that Copernicus intended his reform of astronomy to contribute to astrology. The title, however, was the subject of some controversy. Rheticus objected to the addition of 'orbium caelestium' to the title, yet in his dedicatory letter Copernicus himself uses the expressions 'motuum sphaerarum mundi,' 'motibus sphaerarum orbis,' and 'revolutione orbium coelestium.'[73] Regardless of the reason for Rheticus's objection and Copernicus's genuine intention, the title, as Westman emphasised, is unusual for an astronomical treatise. The term 'revolutio' is standard

---

[69] The letter is edited in Copernicus (1994, no. 95, 186–188). In an appendix Juste cites the translation by Swerdlow (2012, 367–368) (the Appendix to that article, 16–17, was supplemental).

[70] The letter appears in Copernicus (1994, 184–185).

[71] Juste (2023/2027).

[72] Juste (2023/2027).

[73] Copernicus (2015, 7–8).

and typical in astrology, however, referring either to general astrology or the revolution of years that often accompanied a nativity horoscope, analysing the positions of the planets on the anniversary of the individual's birth.[74] I know of only one other example prior to Copernicus, namely, an astrological work of Zahel (or Saul) ben Bishr translated into Latin by Hermann of Carinthia as *de revolutionibus*.[75] For Westman, the term also connoted the relation between periods of revolution and the proper ordering of the planets, or as he expressed it:

> Copernicus's central astronomical argument was that the sidereal return times of the planets with respect to a fixed, central Sun held the key to their true ordering.[76]

As we saw from the Wapowski and Apelt letters, the astrological relevance seems to have been taken for granted. Some experts maintain that the order and relative distances of the planets were irrelevant in constructing horoscopes,[77] yet the initial assignment of qualities to the planets was partly dependent on their order relative to the Sun and their relative distances from each other, in particular Mercury and Venus. In constructing a horoscope, then, a heliocentric ordering would affect the interpretation of the planetary aspects and the earthly houses in which they appear according to their qualitative characteristics. Copernicus, it is true, says nothing about this possibility, unless we see it as implicit in the choice of *Revolutions* as the title of his work.[78]

## 6.7 Conclusion

Now that we have strong reasons for believing that Copernicus was not only not indifferent to astrology but actively practised and promoted it, the Copernican question about the relation between astrology and the heliocentric theory must be reconsidered. Did the lack of consensus among geocentric astronomers about the ordering of the planets motivate Copernicus to find a solution in order to reform astrology, or did he reach his solution as a result of a strictly astronomical critique of geocentric astronomy, and recognized subsequently that the theory and bits observational predictions impacted the practice of astrology? With a different ordering of the planets, the qualitative characteristics assigned to the planets in a geocentric arrangement would

---

[74] Newman and Grafton (2001, 11).

[75] See Tester (1987, 152–153). A work with similar title is Abū Ma'shar's *On the Revolutions of the Years of Nativities* in Abū Ma'shar (2019). See also Bezza (2015, 8–9 and 35–39).

[76] Westman (2011, 45). As Juste (2023/2027) notes, his footnote 25: "A revolution of the nativity is a horoscope cast for the exact time of the return of the Sun to the position it occupied in the nativity, which allows making predictions for one year, from one anniversary to the next." Compare Westman (2011, 135).

[77] Swerdlow (2012, 353–378; 2013, 384–385) for example.

[78] I set aside here discussion of Rheticus's astrological manifesto in *Narratio prima*, because there remains some controversy over Copernicus's reaction to it, a subject that requires more space than available here.

require some adjustment, but it seems as if astrologers continued to focus on the aspects, and we do not know what effect, if any, the change in planetary order had on astrological predictions and practice.

With the available evidence, it seems, particularly from what we know of the exchange between Bernard Wapowski's claims about Copernicus's lost almanac, that the second is likelier.[79] But we still know too little about the content of astrological doctrine and instruction at Cracow in the 1490s to dismiss the first, Westman's hypothesis. The purpose of my paper is to encourage the study of still unedited texts that likely provide evidence of the lectures that Copernicus attended. Those lectures may provide evidence in support of Westman's hypothesis. Did astrologers, in fact, worry about the ordering of Mercury and Venus, or did they just accept Ptolemy's ordering? Until the texts are examined, we cannot resolve the philosophical tension that this paper evokes. That is to say, the tension has not been resolved, but my point is that we cannot resolve it until the relevant texts have been studied. To try to resolve the tension now would be premature and possibly prejudice or close off an open-minded inquiry.

Another source of tension was that between astrological art and astrological science. Everyone recognised the artful dimension of astrological practice. Indeed, it was often cited to explain widely divergent interpretations of the same facts. Astrology was a competitive industry with astrologers and their publishers attacking others for not just incompetence but also poor judgment. Even practitioners questioned the scientific status of astrology. Ludwik Antoni Birkenmajer has taken a letter by Jerome Salius Faventinus to Domenico Maria Novara as evidence of Novara's doubts about astrological science.[80] It seems that much of the effort to improve the accuracy of astronomical observations was an attempt to respond to such doubts and critiques. Copernicus seems to have shared these opinions, supporting his critique of geocentric astronomy. Again the sources need to be examined to see if such a tension was widespread and what remedies were advocated.

It has taken centuries for scholars to consider seriously Copernicus's astrological practice. After the efforts of Robert Westman, there are respectable reasons to pursue his questions and suggestions further. Westman revived the discussion of Copernicus's astrology, and resumed a long-delayed revolution in our understanding of Copernicus. We conclude by calling on experts on observations, meteorology, and medical prescriptions to examine them and Copernicus's other annotations and correspondence in the light of fifteenth–sixteenth-century astrology. Beyond these we have noted the existence of many texts by Cracow scholars on Ptolemy's *Tetrabiblos*, none of which has been edited. The field is ripe for investigation by experts to expand our knowledge of the technical details that Cracow masters of the late fifteenth century taught their students, Copernicus among them, and accordingly to expand on Copernicus's astrological practice in Varmia in the sixteenth century. If Copernicus's astrological activity does not constitute a revolution in our understanding, then it at least marks a major transformation in the historiography of Copernican studies.

---

[79] See Juste (2023/2027).

[80] See Birkenmajer (1990, 437–439); Goddu, (2025, 499–503).

This essay adds some additional detail to the evidence we possess for Copernicus's astrological background in Cracow. Its purpose is to encourage more detailed and concrete investigations of the documents, treatises, and lectures of the 1490s and their possible relation to Copernicus's activity. The concrete evidence of Copernicus's astrological activity is slim by comparison with his strictly astronomical work, but it is not negligible, and it is only now beginning to be examined thoroughly for its technical contents. Astrology was primarily a mathematical discipline dependent on astronomy with a physical interpretation derived from Ptolemy's *Tetrabiblos* and the meteorological and medical tradition. This means that Copernicus's concentration on astronomy absent any reference to astrology does not mean that he considered astrology irrelevant. Astronomical reform would entail astrological reform, an understanding commonly shared by his contemporaries.[81] The discipline of astrology also belonged to the category of the artful, requiring choices, judgment, and the ability to adapt tables and interpret the relation of celestial qualities with terrestrial objects. It is also clear that from an institutional perspective the curriculum in astronomy at Cracow from 1459 on served the curriculum in astrology, and astrology, in turn, served the medical faculty and medical practice.

## References


Abū Ma'shar. 2019. On the Revolutions of the Years of Nativities. In *Persian Nativities III: On Solar Revolutions*, translated and edited by Benjamin N. Dykes. Minneapolis: Cazimi.

Akopyan, Ovanes. 2018. 'Princeps aliorum' and His Followers: Giovanni Pico della Mirandola on the 'Astrological Tradition' in the *Disputationes adversus astrologiam divinatricem*. *Renaissance Studies* 32 (4): 547–564.

Albertus de Brudzewo. 1900. *Commentariolum super Theoricas novas planetarum Georgii Purbachii*. Edited by Ludwik Antoni Birkenmajer. Cracow: Jagiellonian University.

Bezza, Giuseppe. 2014. Representation of the Skies and the Astrological Chart. In *A Companion to Astrology in the Renaissance*, edited by Brendan Dooley, 59–86. Leiden: Brill.

Bezza, Giuseppe. 2015. Saturn-Jupiter Conjunctions and General Astrology: Ptolemy, Abū Ma'shar and Their Commentators. In *From Māshā'allāh to Kepler: Theory and Practice in Medieval and Renaissance Astrology*, edited by Charles Burnett and Dorian Gieseler Greenbaum, 5–48. Ceredigion: Sophia Centre.

Birkenmajer, Ludwik Antoni. 1900. Mikołaj Kopernik. Cracow: Academy of Arts. English edition: Birkenmajer, Ludwik Antoni. 2023. Nicolaus Copernicus. Translated by André Goddu. Toruń: Nicolaus Copernicus University Press.

Biskup, Marian. 1973. Regesta Copernicana. Wrocław: Polish Academy of Sciences.

Blumenthal, Geoffrey. 2014. Copernicus's Development in Context: Politics, Astrology, Cosmology, and a Prince-Bishopric. *Science in Context* 27 (1): 1–32.

Bowden, Mary Ellen. 1990. Review of *A History of Western Astrology* by Jim Tester. *Isis* 81 (4): 754–756.

Cohen, H. Floris, and Noortje Jacobs. 2024. On Exemplary Individuals and the Advantages of Teamwork. *Isis* 115 (3): 593–610.


[81] See Kremer (2006, 226), however, for some qualifications about astrologers' attention to Copernicus's calculations rather than his hypotheses. See also, however, Kremer (2009) on Kepler's situating the discussion in the context of theory and cosmographical system.

Copernicus, Nicholas. 1978. *On the Revolutions.* Translated by Edward Rosen. Baltimore: Johns Hopkins University Press.
Copernicus, Nicholas. 1994. *Nicolaus Copernicus Gesamtausgabe.* Vol. 6/1: *Documenta Copernicana.* Edited by Andreas Kühne. Berlin: Akademie Verlag.
Copernicus, Nicholas. 2015. *De revolutionibus orbium coelestium.* Vol. 2. Edited by Michel-Pierre Lerner, Alain-Philippe Segonds, Jean-Pierre Verdet, and Concetta Luna. Paris: Les Belles Lettres.
Crowe, Michael J. 1990. *Theories of the World from Antiquity to the Copernican Revolution.* New York: Dover.
Czartoryski, Paweł. 1978. The Library of Copernicus. In *Science and History. Studies in Honor of Edward Rosen*, edited by Erna Hilfstein, Paweł Czartoryski, and Frank G. Grande, 355–396. Studia Copernicana 16. Wrocław: Ossolineum.
Dobrzycki, Jerzy. 1975. Mikołaj Kopernik. In *Historia astronomii w Polsce*, vol. 1, edited by Jerzy Dobrzycki, Mieczysław Markowski, and Tadeusz Przypkowski, 127–156. Wrocław: Ossolineum.
Eade, John Christopher. 1984. *The Forgotten Sky.* Oxford: Clarendon Press.
Gingerich, Owen. 2005. *The Book Nobody Read: Chasing the Revolutions of Nicolaus Copernicus.* London: William Heinemann.
Goddu, André. 2004. Copernicus's Annotations: Revisions of Czartoryski's 'Copernicana.' *Scriptorium* 58: 202–236.
Goddu, André. 2009. Copernicus's Mereological Vision of the Universe. *Early Science and Medicine* 14: 316–339.
Goddu, André. 2010. *Copernicus and the Aristotelian Tradition.* Leiden: Brill.
Goddu, André. 2025. The Teaching of Astrology and Astronomy at Cracow in the Fifteenth Century. In Copernicus aus Preußen – Kosmos, Wissen, Politik, edited by Marie-Luise Heckmann, Andreas Kühne, and Sven Tode, 55–71, Hamburg: Nicolaus-Copernicus-Verlag.
Granada, Miguel. 2012. Planetary Order in the Long 16th Century. *Journal for the History of Astronomy* 43: 239–243.
Hartmann, Stefan. 1973. Studien zur Schrift des Nicolaus Copernicus. Ein Beitrag zu Schriftgeschichte des 16. Jahrhunderts. *Zeitschrift für Ostforschung* 22: 1–43.
Hilfstein, Erna. 1980. *Starowolski's Biographies of Copernicus.* Studia Copernicana 21. Wrocław: Ossolineum.
Jan z Głogowa. 1985. *Persuasio brevis quomodo astrologiae studium religioni christiane non est adversum.* Edited by Małgorzata H. Malewicz. *Studia Mediewistyczne* 24 (1): 153–175.
Johannes Pico della Mirandula. 1496. *Disputationes adversus astrologiam divinatricem.* Bologna: Benedictus Hectoris.
Juste, David. 2016. Latin manuscripts. Ptolemaeus arabus et latinus. https://ptolemaeus.badw.de/start. Accessed 9 February 2025.
Juste, David. 2023/2027. Copernicus's Astrological Expertise. Paper presented at a conference in Toruń in September 2023. A revised version is scheduled to appear in Studia Copernicana 47. Turnhout: Brepols, 2027, forthcoming.
Juste, David. n.d. Description of Cracow, Biblioteca Jagellonica, Ms. 2703, Ptolemaeus arabus et latinus. Manuscripts. http://ptolemaeus.badw.de/ms/766. Updated 1 November 2023.
Karliński, Franciszek. 1873. *Żywot Koperniki i jego naukowe zasługi.* Cracow: Jagiellonian University.
Keyser, Paul. 2018. Introduction. In *The Oxford Handbook of Science and Medicine in the Classical World*, edited by Paul T. Keyser and John Scarborough, 1–6. Oxford: Oxford University Press.
Kremer, Richard L. 2006. Copernicus Among the Astrologers: A Preliminary Study. In *Astronomy as a Model for the Sciences in Early Modern Times*, edited by Menso Folkerts and Andreas Kühne, 225–252. Algorismus 59. Augsburg: Dr. Erwin Rauner Verlag.
Kremer, Richard L. 2009. Kepler and the Graz Calendar Makers: Computational Foundations for Astrological Prognostication. In *Johannes Kepler: From Tübingen to Zagań*, edited by Richard

L. Kremer and Jarosław Włodarczyk, 77–100. Studia Copernicana 42. Warsaw: Institute for the History of Science, Polish Academy of Sciences.
Lemay, Richard. 1978. The Late Medieval School at Cracow and the Copernican System. In *Science and History. Studies in Honor of Edward Rosen*, edited by Erna Hilfstein, Paweł Czartoryski, and Frank G. Grande, 337–354. Studia Copernicana 16. Wrocław: Ossolineum.
Lerch, Andreas. 2015. *Scientia astrologiae, Der Diskurs über die Wissenschaftlichkeit der Astrologie und die lateinische Lehrbücher 1470–1610*. Leipzig: AVA Akademische Verlagsanstalt.
*Liber diligentiarum Facultatis Artisticae Universitatis Cracoviensis. Pars I (1487–1563)*. 1886. Edited by Władysław Wisłocki. Archiwum do dziejów literatury i oświaty w Polsce 4. Cracow: Sumptibus Academiae Litterarum.
Markowski, Mieczysław. 1975a. Powstanie pełnej szkoły astronomicznej w Krakowie. In *Historia astronomii w Polsce*, vol. 1, edited by Jerzy Dobrzycki, Mieczysław Markowski, and Tadeusz Przypkowski, 87–106. Wrocław: Ossolineum.
Markowski, Mieczysław. 1975b. Szczyt rozkwitu i międzynarodowego promieniowania krakowskiej szkoły astronomicznej. In *Historia astronomii w Polsce*, vol. 1, edited by Jerzy Dobrzycki, Mieczysław Markowski, and Tadeusz Przypkowski, 107–126. Wrocław: Ossolineum.
Moraw, Peter. 2008. *Gesammelte Beiträge zur Deutschen und Europäischen Universitätsgeschichte: Strukturen - Personen - Entwicklungen*. Education and Society in the Middle Ages 31. Leiden: Brill.
Mikulski, Krzysztof. 2023. *Nicolaus Copernicus. Social Milieu, Background, and Youth*. Toruń: Nicolaus Copernicus University Press.
Newman, William R., and Anthony Grafton. 2001. Introduction: The Problematic Status of Astrology and Alchemy in Premodern Europe. In *Secrets of Nature: Astrology and Alchemy in Early Modern Europe*, edited by William Newman and Anthony Grafton, 1–37. Cambridge, MA: MIT.
North, John. 1986. *Horoscopes and History*. London: The Warburg Institute/University of London.
Pedersen, Olaf. 1962. The Theorica Planetarum-Literature of the Middle Ages. *Classica et mediaevalia* 23: 225–232.
Pedersen, Olaf. 2011. *A Survey of the Almagest*. New York: Springer.
Prowe, Leopold. 1883–1884. *Nicolaus Coppernicus*. 2 vols. Berlin: Weidmannsche Buchhandlung. Repr. Osnabrück, 1967.
Ptolemy, Claudius. 1940. *Tetrabiblos*. Translated by F. E. Robbins. Cambridge, MA: Harvard University.
Rosen, Edward. 1984. *Copernicus and the Scientific Revolution*. Malabar, FL: Robert Krieger.
Rosińska, Grażyna. 1984. *Scientific Writings and Astronomical Tables in Cracow (XIVth–XVIth Centuries)*. Studia Copernicana 22. Wrocław: Ossolineum.
Rosińska, Grażyna. 2002. Krakowski księgozbiór Mikołaja Kopernika w kodeksach *Copernicana 4* i *Copernicana 6* Biblioteki Uniwersyteckiej w Uppsali. *Res Historica* 13: 105–123.
Rosińska, Grażyna. 2006. "Mathematics for Astronomy" at Universities in Copernicus' Time: Modern Attitudes Toward Ancient Problems. In *Universities and Science in the Early Modern Period*, edited by Mordechai Feingold and Victor Navarro-Brotons, 9–28. Dordrecht: Springer.
Rutkin, H. Darrel. 2019. *Sapientia Astrologica: Astrology, Magic and Natural Knowledge, ca. 1250–1800*. Cham: Springer.
Swerdlow, Noel M. 2012. Copernicus and Astrology, with an Appendix of Translations of Primary Sources. *Perspectives on Science* 20 (3): 353–378; Appendix, 1–23. http://www.mitpressjournals.org/doi/pdf/10.1162/POSC_a_00071. Accessed 26 September 2024.
Swerdlow, Noel M. 2013. On Professor Westman's Reply to *Copernicus and Astrology* with an Appendix of Translations of Additional Primary Sources. *Perspectives on Science* 21 (3): 384–385; Appendix, 1–20. http://www.mitpressjournals.org/doi/suppl/10.1162/POSC_c_00105. Accessed 26 September 2024.
Swerdlow, Noel M., and Otto Neugebauer. 1984. *Mathematical Astronomy in Copernicus's* De Revolutionibus. 2 parts. New York: Springer.

Tessicini, Dario. 2013. Astrology, Heliocentrism and the Copernican Question. *Galilaeana: Journal of Galilean Studies* 10: 219–236.

Tester, S. Jim. 1987. *A History of Western Astrology*. Woodbridge: Boydell Press.

Thorndike, Lynn. 1934–1941. *A History of Magic and Experimental Science*, vols. 4–5. New York: Columbia University Press.

Vanden Broecke, Steven. 2003. *The Limits of Influence: Pico, Louvain, and the Crisis of Renaissance Astrology*. Leiden: Brill.

Wasiutyński, Jeremi. 1938. *Kopernik. Twórca nowego nieba*. Warsaw: J. Przeworski.

Wasiutyński, Jeremi. 1963. Uwagi o niektórych kopernikanach szwedzkich. *Studia i Materiały z Dziejów Nauki Polskiej*, Seria C 7: 65–84.

Westman, Robert S. 2011. *The Copernican Question: Prognostication, Skepticism, and Celestial Order*. Berkeley: University of California Press.

Westman, Robert S. 2013. The Copernican Question Revisited: A Reply to Noel Swerdlow and John Heilbron. *Perspectives on Science* 21: 100–136.

Westman, Robert S. 2017. *Copernicus and the Astrologers: Dibner Library Lecture, 12 December 2013*. Washington, DC: Smithsonian Libraries.

Westman, Robert S. 2026. Copernicus and Astrology: Remarks on the State of the Question. *Studia Copernicana*. New series, Vol. 1. Turnhout: Brepols, forthcoming.

Whitfield, Peter. 2001. *Astrology: A History*. New York: Harry N. Abrams.

**André Goddu** is emeritus professor of astronomy and physics at Stonehill College, Easton, MA, USA. His most recent book is a translation of Ludwik Antoni Birkenmajer's Mikołaj Kopernik (Cracow, 1900): *Nicolaus Copernicus, Part One, Studies on Copernicus's Works and Biographical Materials* (Toruń: Wydawnictwo Naukowe Uniwersytetu Mikołaja Kopernika, 2023).

# Chapter 7
# Early Seventeenth-Century Atomists on Heat and Fire

**Antonio Clericuzio**

**Abstract** In the early seventeenth century, new empirical knowledge together with the revival of ancient atomism prompted philosophers and physicians to reconceptualise theories of heat and fire in corpuscular terms. Two atomists, i.e., Isaac Beeckman and Sébastien Basson, abandoned the Peripatetic doctrine of forms and qualities and advanced a kinetic theory of heat, whereby the shape, size, and motion of material particles explained heat and fire. Galilei, who adopted a mechanical theory of qualities, argued (*The Assayer*) that heat is merely a name for the action of minute corpuscles that strike the sense organs. Heat is caused by tiny particles entering our flesh; depending on their quantity and speed, they create either pleasant warmth or painful burning. Galileo explored the relationship between light and heat, and maintained that light is made of extremely subtle, almost immaterial atoms that can expand infinitely. Benedetto Castelli, Galileo's pupil and collaborator, carried out experimental research on the relationship between heat and light and concluded that a brick's absorption of heat from the sun's rays depended on the colour of its surface. Castelli's experiment was repeated by Robert Boyle, who thoroughly explored this topic in his *History of Colours* (1664).

**Keywords** Heat · Fire · Atomism · Light · Motion

## 7.1 Introduction

Since antiquity, heat and fire have attracted the attention of philosophers and physicians. In the *Timaeus* (56a–b), Plato offered a geometrical account of fire by associating it with the tetrahedron—the simplest and sharpest of the regular solids—thus explaining its piercing properties. Aristotle made fire one of the four elements, although, he maintained, the true elements of the natural world were not earth, water, air, and fire, but the four chief physical opposites: Hot and Cold, Dry and Wet. It

A. Clericuzio (✉)
Roma Tre, Rome, Italy
e-mail: antonio.clericuzio@uniroma3.it



E. Jung et al. (eds.), *The Forgotten Giants: Overlooked Revolutions in Natural Philosophy*, Studies in History and Philosophy of Science 70,
https://doi.org/10.1007/978-3-032-34008-5_7

is from the combination of these opposing principles that the four elemental bodies arise.[1] Aristotle's student Theophrastus (*De igne*) denied fire the status of a simple body. By emphasising the difference between fire and the other material principles (air, water, earth), he prepared the ground for a discussion concerning the relation of the material principle of fire to its peers.[2] Lucretius, for his part, rejected the doctrine of the four elements, arguing instead that solid and unchangeable atoms endowed solely with shape, size, and motion constituted the ultimate components of bodies. Collisions, motions, order, positions, and shapes of atoms generate fire, but, when their arrangement changes, the nature of what they produce changes as well (*De rerum natura*, Book 1, 684–689). As we shall see, early seventeenth-century atomists adopted and developed Lucretius' outlook that the *texture of atoms* was responsible for the production of the qualities of bodies.

During the Renaissance, the nature and status of fire became a subject of intense debate. Cardano, among others, denied that fire was an element, reducing it instead to a compound. He argued that an element is that which does not require nourishment for its existence—and this, he contended, is not the case with fire. Cardano, who aimed at reforming Aristotelian natural philosophy along empirical lines, maintained that fire, owing to its subtlety, passed through the pores of bodies, separated their parts, burned what was separated, and converted it into its own substance.[3] Cardano maintained that celestial heat, being no elemental quality, is diffused throughout the world and is the source of all terrestrial heat. It is the efficient cause of all generation and the active principle in the formation of natural bodies, including minerals and metals.[4]

In the early decades of the seventeenth century, certain natural philosophers abandoned the notion of fire as an element endowed with intrinsic heat and instead reconceptualised it as a physico-chemical process resulting from the motions of the particles of matter.[5] The aim of this paper is to tackle the early modern emergence of the understanding of fire and heat in terms of motion of corpuscles. This process originated from the interplay of different factors, including the recovery of ancient atomism, the technology of fire as developed by alchemists and metallurgists throughout the Middle Ages. Moreover, Galenic and Avicennian pharmacology, especially the study of drug powers in terms of the degrees of the primary qualities (i.e., the so-called *compositio medicamentorum*), and the related investigation of the property of the mixture underpinned the empirical investigation of heat.[6] Hence, it is hardly surprising that physicians played a relevant part in the reinterpretation of heat and fire along experimental and corpuscular lines. A leading figure in this development was Santorio Santori, a renowned physician who applied quantitative

[1] Kahn 1960, 120.

[2] van Raalte 2010, 47.

[3] Cardano 1550, 27–28; Ingegno 1980, 224–226.

[4] Cardano 1550, 325. Taape 2025. Telesio denied the elementary nature of fire too and adopted Cardano's view of celestial heat as the active substance. For Bernardino Telesio see Hirai 2019.

[5] See Salvia 2025, 51.

[6] Bigotti 2017 for Santorio. See also Beeckman 1939–1953, 1:130–131, 134.

methods to medicine and supported the corpuscular theory of matter.[7] He is credited with 'inventing' the medical thermoscope, an instrument for measuring relative variations in heat.[8]

The first part of the present paper explores early seventeenth-century theories and experiments on heat and cold, focussing on Isaac Beeckman and Sébastien Basson, two physicians who expounded atomism and dispensed with the Aristotelian doctrine of forms and qualities. As Lüthy pointed out, "For Basson, like for many neo-atomists, fire and the phenomena grouped around it—rarefaction, heat, distillation, inflammability, and light—provided the most decisive evidence against Aristotle."[9] The second section is devoted to Galileo's research of heat and fire, paying special attention to his views of the relationships between fire and light. Finally, the paper takes into account Castelli's little-known experimental investigation of the differential absorption of light by colours and its mechanical explanation of the relationship of light to heat.

## 7.2 Early Corpuscular Theories of Heat: Beeckman and Basson

An advocate of atomism, the Dutch physician and natural philosopher Isaac Beeckman (1588–1637) pursued the mechanical explanation of natural phenomena and of sensible qualities, notably heat. Isaac Beeckman was a physician, a craftsman and natural philosopher who carried out experimental investigations of a wide range of topics, such as mechanics, acoustics, optics, magnetism, hydrostatics, medicine, and developed a mechanical theory of matter. He published virtually nothing during his lifetime (only his medical dissertation saw the light in 1618), yet, he played a relevant part in seventeenth-century science through his correspondence and his contacts with scientific practitioners and natural philosophers, including Mersenne, Descartes and Gassendi. Besides his letters, evidence for his natural philosophy is to be found in the diary he kept from 1604 to 1634, which Cornelis de Waard published as *Journal tenu par Isaac Beeckman de 1604 à 1634*.[10] Beeckman did not expound a system of natural philosophy, but resorted to atoms and void to explain a wide range of natural phenomena and the sensible qualities of bodies too. All the differences

---

[7] For the biography of Santorio, see Barry & Bigotti 2022, 1–63. For Santorio's corpuscular theory of matter see Bigotti 2022, 65–102.

[8] Salvia 2025, 57–62.

[9] Lüthy 1997, 13.

[10] Gemelli 2002; Van Berkel 2013. Excerpts from his *Journal* compiled by his brother Abraham, were published posthumously in 1644 as *Mathematico-physicarum Meditationum, Quaestionum, solutionum Centuria.*

in natural bodies ultimately derive from the motions of atoms, namely, from the so-called secondary qualities, shape and order.[11] He placed the different properties of bodies in the figures and motion (or rest) of atoms, as well as in the pores or microscopic empty spaces between corpuscles of matter.[12] A devout Calvinist, Beeckman ruled out the Epicurean and Lucretian doctrine of the fortuitous concourse of atoms by maintaining that God created atoms and ruled the physical world according to a providential design.[13] In 1617 Beeckman surmised that there were four kinds of atoms, corresponding to the four elements, all made of the same matter and having different shapes.[14] The four primary qualities of Peripatetic philosophy are perceived by touch, but are understood by intellect only if they are referred to the shapes and motion of the corpuscles and of pores. The nature of heat and cold appears to consist in motion, so that the fast motion of corpuscles produces heat, very slow motion is responsible for cold. He took issue with the Peripatetic theory of elements, claiming that they are no simple substances as they can be broken down into "suas atomos."[15] As he wrote in his diary for 1618, "The nature of the moist and the dry consists in the shape of atoms. Dry atoms are pointed, while moist atoms approach roundness and are more spherical than pointed."[16] In 1620 he saw sensible qualities as subjective experience, therefore adopting a mechanistic account that was similar to the one Galileo articulated in *Il Saggiatore*:

> Heat and cold, etc., are merely dispositions (*affectiones*) of the sentient body impressed by the variety of corpuscles. [...] There are four qualities in the four elements: for atoms (*primordia*), combined in one way, constitute fire; in another, a different element; by the varied figure of their conjunction, which affects us in four ways; and for this reason qualities are said to be in the elements, since they are nothing but the ingress of diverse figures into our body, in the same way that a knife affects us one way, a stick another.[17]

---

[11] Beeckman 1939–1953, 1:129: "Attamen quid obstat quominus ea. nihilominus consisteret in secundis dictis qualitatibus, id est in diversarum figurarum particulis, diverso modo ad invicem dispositis?"

[12] "Bodies have two attributes: first form, second motion, and rest. Once moved, they never came to rest, unless impeded by each other. Therefore, combined with the interstitial vacuum, these constitute matter and form of the composite bodies in heaven and Earth." (Beeckman 1939–1953, 1:132). Van Berkel 2013, 218. When Beeckman spoke of form he meant the geometrical form.

[13] Van Berkel 2013, 143.

[14] Beeckman 1939–1953, 1:152–153.

[15] Beeckman 1939–1953, 2:82: "Nec tamen necesse est hasce essentias ex elementis mixtis componi, cum ex ipsis atomis immediate multa possint nasci, imo ipsa elementa inter se mixta possint solvi in suas atomos naturamque elementi deponere, ita ut compositum potius ex indivisibilibus principijs quam ex elementis constet."

[16] Beeckman 1939–1953, 1:152, 216.

[17] Beeckman 1939–1953, 2:83: "Calor autem et frigus etc. sunt duntaxat affectiones corporis sentientis a corpusculorum varietate impressae [...]; 2:85: Quatuor autem qualitates sunt in quatuor elementis: primordia enim hoc pacto conjuncta constituunt ignem, alio pacto aliud elementum; qua primordiorum conjunctio suâ figurá variâ quadrifariam nos afficit, ob idque dicuntur in elementis esse qualitates, cùm nihil sint nisi diversarum figurarum in corpus nostrum ingressüs, eo modo quo nos aliter afficit culter, aliter baculus."

According to Beeckman, cold is just lack of heat.[18] Along with Lucretius, he stated, the same atoms are fire when they move rapidly, but when at rest they are inflammable substances. Sulphur and fatty substances are denser than air, but become lighter than air by their movement alone.[19] The effects of heat and cold on metals are explained by Beeckman along mechanical lines, namely, by resorting to the motion and size of particles and to the size and arrangement of pores. He addressed this topic in a note to Vitruvius's statement to be found in *Ten Books on Architecture* (book 1, Chap. 4, Sect. 3). In it, Vitruvius maintained that heat makes iron softer and more malleable. After dipping it while still soft and white-hot into cold water, it becomes hard again and regains its former quality. Beeckman provided the following account:

> [iron softens] because heat spreads evenly throughout the iron, it dislocates all its particles equally, makes the pores equal, and separates all the homogeneous particles from each other. [...] If, however, it [i.e., hot iron] is quenched in cold water, cold forces the fire to issue in a dense stream; it dislodges the particles of iron and makes the pores larger, since it is easier to pass through one opening twice as wide than through two half-open ones. For things in motion, being joined together, move more powerfully and are not so easily arrested, as I said before about smoke [...] Therefore, when larger pores have formed in the iron and the whole body remains of the same capacity, many particles necessarily come together to make very small pores there, and thus they exhibit great hardness. For the smaller the pores, the harder a thing is; indeed, that in which there are no pores, as in its primordial state, cannot be broken and is the hardest.[20]

As the passage quoted above shows, his explanation of hardness was strictly mechanical. Beeckman argued that hardness depends on the number of pores in a body: the fewer the pores, the harder the body. Atoms are perfectly hard and solid because they contain no pores.[21]

A physician, Beeckman dealt extensively with qualities and temperaments, and also investigated the composition and powers of remedies along Galenic and mechanical lines. He offered a corpuscular interpretation of innate heat, which he regarded as consisting in the element of fire applied to a combustible bodily material, the "radical moisture." He described this moisture as a flammable and combustible substance, similar to sulphur.[22]

The kinetic theory of heat was held by Sébastien Basson, a physician, schoolmaster, and natural philosopher from Lorraine. Basson authored the *Philosophia naturalis* (1621), a lengthy textbook that refuted Aristotelian philosophy by appealing to theories he regarded as the legacy of the ancient philosophers. As we read in the subtitle of the work, the author intends to restore the "abstrusa veterum physiologia" (the abstruse natural philosophy of the ancients). Basson did not provide a detailed

---

[18] Beeckman 1939–1953, 2:385.

[19] Beeckman 1939–1953, 3:138: "ignis autem, ut ante sæpe, non est nisi motus talium atomorum, quæ sua quiete constituunt sulphur, pinguia, inflammabilia, eaque sunt magis densa quam aer, ita ut solo motu aere sint leviora."

[20] Beeckman 1939–1953, 2:384.

[21] Beeckman 1939–1953, 2:384: "Quo enim pori minores, eo durius aliquid est; imo in quo nulli pori, ut in primordijs, ea. frangi nequeunt et durissima sunt."

[22] Moreau 2018, 353–360. On innate heat see Mendelsohn 1964.

account of the properties of atoms; rather, he advanced a corpuscular interpretation of the four elements and of the chemical principles. When he discussed the atoms of the elements, he assigned a specific shape only to the atoms of fire, which were sharp and thus capable of penetrating bodies. Moreover, fire particles are thinner than those of the other elements and move faster.

For Basson, matter was not inert but endowed with powers—namely attraction, repulsion, sympathies, and antipathies—ultimately deriving from the universal spirit (*spiritus*). *Spiritus,* which is the source of motion of corpuscles, constitutes the direct instrument of God's action in the physical world. By imbibing the fiery atoms, the spirit drives them into very swift motion, while the purest fire resides in the stars.[23]

Like Beeckman, Basson dealt extensively with sensible qualities—most notably with heat—and argued that natural phenomena and their qualities originate solely from the size and local motion of particles of matter. The four elements and the five chemical principles are higher-level compounds that act upon us not for their presence, but because of their action on our senses, while the smallest indivisible particles of matter do not.[24]As Basson put it, "Heat, cold, humidity, and dryness, which are called primary qualities, derive from the motion of atoms."[25] The perception of heat is explained in terms of moving particles and pores. Heat produces a pleasant sensation in us if the motion of the fiery particles is not violent and therefore pleasantly excites our heat, but, if it is violent, it causes pain because it separates the parts of our body and causes the particles of fire in our body to escape rapidly through the pores.[26] The motion of fiery corpuscles serves as the *explanans* for a variety of chemical reactions. The corpuscles of fire (*igniculi*) break apart the bodies upon which they act by virtue of their shape, being sharp corpuscles akin to spikes[27] Having rejected the Aristotelian doctrine that heat associates things of the same kind (*De gen. et corrupt.* 329b, 26–299) he argued that "the nature of heat, that is, the power of fire, and its action is not to bring together like things, but to divide and cut things thanks to its pointed particles."[28] No substance takes fire spontaneously; rather, ignition results from the action of external bodies—most notably sun rays or fires—that extract and liberate the fiery particles (*igniculi*) contained within bodies. For Basson, heating

[23] Basson 1621, 434–435. Zanier 1977, 94–102.

[24] Basson 1621, 126.

[25] "Praecedenti libro egimus de illo motu quem localem esse nemo negat, praesenti statuimus ostendere quam sit verum quod affirmavimus ex Veterum disciplina, solo motu locali particularum, ex quibus res queque fit, omnes mutationes fieri, quas diversissimas res quelibet patitur. Id autem sufficienter fuerit declaratum, si calorem, frigus, humiditatem, siccitatem, quas primas qualitates vocant, ex tali motu provenire probaverimus." (Basson 1621, 430)

[26] Basson 1621, 470: "E re tacta calida igniculi exeunt, qui si poros nostros sine violentia subeant, ignem nostrum suaviter excitant. [...] sin vero externus ille ignis vehementior prae iaculorum multitudine partium nostrarum faciat divisionem, dolorem patit. Nec solum inde provenit dolor, sed quia ignis intimus laxior factus est, dum, extrorsum contendit, maiorem quoque eruptione partium dissolutionem facit." Basson's view was analogous to the one Galilei espoused in *Discorso delle comete* (1619), Galilei 1968, 6:56, see below.

[27] Basson 1621, 109.

[28] Basson 1621, 463.

was essentially the action of drawing forth these fiery particles from a body.[29] When fiery particles therein contained can escape from the combustible body (for instance from wood)—due to the opening of the pores—they bring about flame. Fermentation brings about heat because fire particles move freely in the fermenting substance, whereas an exothermic reaction of quicklime and water occurs because water extracts the fiery particles contained in quicklime. This happens because water dissolves the bonds that limited the motion of fire particles.[30]

As far as cold was concerned, Basson was rather ambivalent. He stated that cold causes pores to contract and repels heat, causes particles to stick together (provided that there is moisture), yet he added that cold was no positive quality, being nothing but want of heat, or the outcome of the fiery parts being compressed in a body.[31] In the second half of the seventeenth century, the status of cold remained a matter of debate, as evidenced by Boyle's objections to Gassendi. The latter had maintained that cold constituted a real and positive quality and had designated nitre as the principle of cold, the *primum frigidum,* insofar as it enclosed *frigoris semina*—atoms of cold configured as tetrahedra. It was their geometrical shape that conferred the capacity to cool.[32] By contrast, Boyle, in *Of the Mechanical Origin of Heat and Cold* (1675), maintained on an experimental basis the privative character of cold, arguing that cold resulted from the privation of motion. Yet he did not rule out that some agent might ultimately be responsible for cold.[33]

## 7.3 Galileo: Atoms, Heat and Fire

In a well-known passage of *The Assayer* Galileo argued that sensible qualities were not objective, being but names, which we attribute to the impressions that particles of matter produce in our sense organs. By contrast, form, size and motion are properties inherent in bodies, objective in that they exist independently of the sentient subject.[34]

---

[29] Basson 1621, 93–106, 129.

[30] Basson 1621, 102–103.

[31] Basson 1621, 464–465, 473: "frigus [...] nihil aliud est quam vel caloris privatio ut in terra; vel ignis in ea. re inclusi compressio."

[32] Gassendi 1649, 324–331.

[33] Clericuzio 2023, 355–356.

[34] Galilei 1957, 274: "Now I say that whenever I conceive any material or corporeal substance, I immediately feel the need to think of it as bounded, and as having this or that shape; as being large or small in relation to other things, and in some specific place at any given time; as being in motion or at rest; as touching or not touching some other body; and as being one in number, or few, or many. From these conditions I cannot separate such a substance by any stretch of my imagination. But that it must be white or red, bitter or sweet, noisy or silent, and of sweet or foul odor, my mind does not feel compelled to bring in as necessary accompaniments. Without the senses as our guides, reason or imagination unaided would probably never arrive at qualities like these. Hence I think that tastes, odors, colors, and so on are no more than mere names so far as the object in which we place them is concerned, and that they reside only in the sensitive body. Hence if the living creature were removed, all these qualities would be wiped away and annihilated." I have slightly revised Stillman

The main goal of Galileo's well-known theory of the origin of qualities was to vindicate the cognitive value of common sensibles, a topic he had addressed in his reply to Giulio Cesare La Galla. It is noteworthy that Paolo Sarpi embraced the analogous conception of the proper sensibles and underscored the cognitive function of the common sensibles.[35] Aristotle distinguished between proper sensibles and common sensibles. Proper sensibles are perceived by a single sense: colour by sight, sound by hearing, odour by smell, flavour by taste, heat, cold, dryness, wetness, hardness, softness, etc. by touch, while common sensibles (motion, rest, number, shape and size) are perceptible by more than one sense at once. According to Aristotle (*De anima* 425a15 and *De Sensu*, 442b5–6), proper sensibles do not cause deception (or at least very seldom), while the common sensibles could occasion it.[36] Galileo endorsed Democritus' views of the properties of bodies, which he may have learned about from Aristotle (*De Gen et Corrupt.*, 316a1; *De Sensu*, 446a20) and Galen 1996 (*De elementis ex Hippocratis sententia libri duo*). As Taylor pointed out, Democritus described the perception of secondary qualities as a response, not to the properties of an individual atom, but to a physical stimulus constituted by a continuous bombardment of atoms.[37]

Galileo's doctrine of qualities entailed the corpuscular theory of matter, which he adopted at an early stage of his scientific career, as attested by *De motu antiquiora*, written between 1589 and 1592.[38] Galileo voiced his support to Democritus' physics (as opposed to Aristotle's) in *Discorso intorno alle cose che stanno in su l'acqua o che in quella si muovono* (1612), where he held the (somehow revised) Archimedean theory of floating bodies, and resorted to the comparison of the specific weights (of the solid to the one of the fluid). He interpreted the physical properties of fluids along corpuscular lines, by stating that fluidity was due to the absence of bonds holding corpuscles together.[39] Furthermore, Galileo claimed that the so-called natural motion of elements could be explained in terms of varying density, namely as the ratio between parts of matter (plenum) and empty spaces to be found in bodies. He spelt out this theory by resorting to Democritus:

> Democritus, who was not content with discussing names, sought to understand what gravity and lightness are, or rather, the cause of downward and upward motion. He therefore made use of plenum and void, attributing void to fire (which therefore moves upwards) and plenum

Drake's translation. I have used the new edition of *Il Saggiatore*, with an excellent commentary, see Galilei 2023, 246–248.

[35] Sarpi 1996, 274: "Tutti li sensibili propri sono cose che da estrinseco passan per li sensori alla conoscitiva e alla memoria, ed eglino si dicono esser nella mente *subiective*: ma li comuni e altre intenzioni, o sensibili o intelligibili, non passan da di fuori, bensì fabbricati sono dalla mente, e l'esser che hanno in lei chiamasi *obiective*, e tutto quello che nella mente così è, dicesi concetto." See Redondi 1985, 537–538.

[36] For the Aristotelian theory of common sensibles, see Owens 1982.

[37] While Democritus claimed that all that exists in the world is atoms (having two basic properties, i.e., size and shape) and the void, Epicurus saw the qualities of macroscopic bodies as real Taylor 1999, 177. See also Furley 1993; Pasnau 2007.

[38] Galluzzi 2011, 11–19.

[39] Galluzzi 2011, 29–54.

> to earth (which therefore moves downwards), while attributing a greater amount of plenum to air than to fire and more void to water than to earth.[40]

Galilei, who adopted a strict mechanical theory of qualities, argued that heat is merely a name for the action of minute corpuscles that strike the sense organs. His view is well summarised in the annotation to a letter from Giovanni Baliani of 1612:

> The heat we feel comes not from the substance, but from the shape, size and motion of the body dissolved into minute parts. To wound and pierce the flesh, it is not necessary for the knife to be made of steel rather than copper, stone, bone or oak; it is enough that it is sharp and cutting: and so, it matters little that the smallest parts into which wax dissolves are of a different substance from those into which iron dissolves in order to generate heat in us, as long as both dissolve into very thin, sharp and mobile parts, that is, capable of penetrating our pores.[41]

As attested by undated manuscript notes (but, as suggested by Galluzzi 2011, related to his correspondence with Giovan Francesco Sagredo of 1615), Galilei explored the perception of heat and cold in terms of motion of corpuscles, and the sizes of pores.[42] It is also noticeable that in the same manuscript Galilei maintained that cold was nothing but want of heat (with: "onde ne segue che il freddo non sia altro che privazione di caldo."[43]

In *Risposta alle opposizioni del S. Lodovico delle Colombe* (1615) Galilei vindicated Democritus' physics against Aristotle's, and explained water boiling in terms of motion of *ignicoli* (i.e., atoms of fire) as follows:

> you will see that as the atoms of fire increase, several congregate in the water and form small globules which rise in great number and escape at the surface.[44]

He saw heat as the outcome of the motion of particles of matter, and interpreted the experiments made with thermoscope in terms of rarefaction and condensation,

---

[40] Galilei 1968, 4:133: "egli [i.e., Democritus], non si contentando del nome solo, aveva voluto più particolarmente dichiarare che cosa fusse la gravità e la leggerezza, cioè la causa dell'andare in giù e dell'ascendere, e aveva introdotto il pieno e 'l vacuo, dando questo al fuoco, per lo quale si movesse in su, e quello alla terra, per lo quale ella discendesse, attribuendo poi all'aria più del fuoco e all'acqua più della terra."

[41] Galilei 1968, 12:475. Redondi 1985, 547.

[42] Galilei 1968, 8:685: "Che gli uomini muoino intirizzati dal freddo, avviene che il freddo ambiente va consumando tutti quegli atomi ignei che trova nelle membra, onde, non v'essendo più il calor naturale, si muore. L'acqua posta in una stanza si trova nella medesima temperie che la stanza dove si pone, partecipando amendue egualmente di atomi ignei. Ma che una mano, che tenuta in aria ti par calda, poi posta nell' acqua si raffredda, questa ne è la cagione, considerandosi e il caldo esterno e l'interno, che, mentre resta in aria, gli atomi ignei suoi proprii hanno luogo d'uscire, che son quelli che cagionano il caldo; ma posta in acqua, le particole d'essa turano e serrano gli aditi onde escono i detti atomi, essendo le parti dell'acqua maggiori delle porosità per le quali essi scappano fuori; il che non avviene nell'aria, trovando il campo libero, come quelli che non son tenuti dalle parti dell'aria, per esser minori de i pori onde erumpunt: essendo che il caldo non sia altro che il contatto e solleticamento di quei atomi calidi, i quali, nello scappar fuora, toccano le membra del corpo.." See Galluzzi 2011, 52.

[43] Galilei 1968, 8:684.

[44] Galilei 1968, 4: 654–655. See Shea 1970, 14; Redondi 1985, 533–534 and Salvia 2025.

a matter he discussed in his correspondence with Giovan Francesco Sagredo in 1615. The thermoscope was composed of a small glass vessel furnished with a very slender tube, approximately fifty centimetres in length. The vessel was first warmed by the hands and subsequently inverted and immersed in a jar containing of wine. As the vessel gradually lost the heat imparted to it, the wine was observed to ascend within the tube. For Galileo, the experiment illustrates how variations in temperature affected air density. When the jar was heated, the air inside expanded, resulting in a decrease in the wine level within the tube. Conversely, as the air cooled, its volume contracted, allowing the water to rise from the lower reservoir into the jar's tube. Galileo accounted for the ascent of wine (or water) by invoking the principle of *fuga vacui*.[45] Furthermore, Galileo held that cold was nothing more than the absence of heat, and thus argued that a body cools as its fiery atoms escape.[46]

It is well-known that Galileo's *The Assayer* originated from his rebuttal of the cometary theory held by the Jesuit Orazio Grassi's (alias Lotario Sarsi). Galileo replied to Sarsi's *De tribus cometis* (1619) in *Discorso delle co*mete (1619), published under the name of his former disciple Mario Guiducci, but largely written by Galileo. In the *Discorso* he took issue with the Aristotelian outlook of heat to be found in *De caelo* (289a 20–35), where heat is seen as the outcome of motion and friction—a view that Grassi held too. According to Galileo, heat arises from the friction and compression of solid bodies with irregular surfaces, which lead to their disintegration into small moving particles.[47] He articulated his view of heat along the following lines:

> The body that produces heat must be dissolved into very small particles, which penetrate the pores of our flesh as they move, and by penetrating them, according as they are few or many, slow or fast, they produce in us by contact either a certain pleasant feeling which we call warmth, or else, by a violent dissolution of the parts, the pain called scorching or burning.[48]

Galileo explored the relationship of heat to light too. In the letter to Monsignor Piero Dini of 1615 he claimed that the calorific spirit coming from the Sun, being highly volatile, thin and fast, was

> something more than light, in that it penetrates and spreads out through all bodily substances, even the densest ones, through which light is not able to penetrate in similar manner" (*qualche cosa di più del lume, poi che ei penetra e si diffonde per tutte le sustanze corporee, ben che densissime, per molte delle quali non così penetra essa luce*).[49]

---

[45] For a detailed account of Galileo's thermoscope, see Salvia 2025.

[46] Galluzzi 2011, 48–52.

[47] Galilei 1968, 6:56: "Ma quando al moto, o alla compressione, ne seguita l'arrotamento della superficie del corpo mobile con altro corpo solido, o lo stropicciamento delle interne parti tra di loro, allora ne segue il calore. E notisi di più, non di qualsivoglino corpi solidi la confricazione produr calore, ma solamente di quelli, che nel fregarsi insieme, amenduni, o almeno uno, si consuma, e, per così dire, si polverizza; che, se, o per essere i corpi sommamente duri, o per esser di superficie terse, e lisce, accadrà, che nello stropicciarsi insieme nulla di loro si stacchi, e consumi, vana sarà ogni fatica per riscaldargli."

[48] Galilei 1968, 6:56. See Galluzzi 2011, 68.

[49] Galilei 1968, 5:301.

Subsequently, in *The Assayer*, he stated that light consisted of very thin and almost incorporeal atoms and therefore was able to indefinitely expand in the universe. He adopted two distinct kinds of atoms. One kind of atoms he saw as units of matter having dimensions and weight, which he called extended minims (*minimi quanti*), the other, namely the indivisible atoms of light, he described as extremely subtle (Galilei surmised they were immaterial), indivisible, imponderable particles. The former kind of atoms move over time (*il moto loro è temporaneo*), the latter have instantaneous motion (*espansione e diffusione istantanea*).[50] Galileo's outlook of the composition of light to be found in *Il Saggiatore* was not his total thought on this subject, however. In the *Discorsi e dimostrazioni matematiche intorno a due nuove scienze* (1638) he maintained that the speed of light was very high, but finite.[51] Galilei also argued that matter was composed of infinite non-extended atoms (*infiniti atomi non quanti*).[52]

Galileo recognised that the complexity of optical phenomena required extreme caution in formulating explanatory theories and was not inclined to formulate hypotheses about the nature or essence of the physical objects under investigation.[53] As Camerota pointed out, Galileo refrained from providing a definition of the nature of light.[54]

## 7.4 Benedetto Castelli on Light and Heat

Benedetto Castelli (1578–1643), a Benedictine monk and Galileo's student and collaborator, conducted experimental investigations of heat and light. He is well known for his important contributions to hydraulics, mechanics, and mathematics.[55] In a letter to Galileo dated 8 May 1612 about sunspots, Castelli outlined his hypothesis

[50] Galilei 2023, 252–253.

[51] Galileo's *Istoria e dimostrazioni intorno alle macchie solari* (Galilei 1968, 5:71–239, esp. 230–231), letter to Dini of 1615 (Galilei 1968, 5:197–305, esp. 301–302), and *Discorsi* (Galilei 1968, 8:89). See Gómez 2013.

[52] Galilei 1968, 8:26. Galluzzi 2011. As Palmerino argued, "Lack of extension, which in *The Assayer* was considered to be the sole distinctive property of the ultimate particles of light, was to become in the *Two New Sciences* the common attribute of all atoms" (2011, 311–312).

[53] Galilei to Liceti, 23 June 1640, in Galilei 1968, 18:208: "[...] I speak of the essence of light, concerning which I have always remained in darkness." In *Istoria e dimostrazioni intorno alle macchie solari* (1613) he articulated his rejection of essentialism along the following lines, "I consider investigating the essence of the nearest elementary substances an undertaking no less impossible and a labor no less vain than that of the most remote and celestial ones." (Galilei 1968, 5:187). See Garber 2013, 14.

[54] As Camerota pointed out, "Il problema di comprendere l'intrinseca costituzione della luce rimarrà sempre vivo nella mente di Galileo. Gli esiti della sua riflessione non approderanno però a risultati definiti." (2025, 173). See also Galluzzi 2011, 122.

[55] For Benedetto Castelli, see De Ferrari 1978; Favaro 1983, 2:738–870; Bucciantini 1992.

on light. He claimed that a luminous body vibrates continuously and emits extremely fine and fast-moving particles, which we call light.[56]

Years later, he wrote an account of his investigations into heat and light in three letters to Galileo, dated 27 June 1637, 9 August 1637 (the latter being a revised version of the former with minor changes), and 15 August 1637. The second and third letters were published as a separate tract in a collection of scientific papers printed posthumously in 1669. Castelli demonstrated experimentally that the absorption of heat from the sun's rays by a brick was dependent on the colour of its surface. As we read in the first letter to Galileo, the experiment was prompted by a discussion he had with one of his students on heat and cold—a discussion that inevitably invoked Galileo's *Il Saggiatore*. Castelli then sent Galileo the following account of the experiment, which showed the differential absorption of heat by surfaces of different colours:

> If a brick were taken and painted on one side half black with ink and half white with lime, and exposed with the painted side to the sun's rays, and left there for half an hour, or a little more or less, it would heat up noticeably more on the black side.[57]

Castelli also tested whether the same result occurred when the brick was exposed to fire:

> What increases the wonder of the reported event is that when the same brick was exposed to the heat of a fire, the black part was just a little hotter than the white part and for this result too I believe that I have an explanation. But I shall inform you of it another time for first the query will be posed to the oracle.[58]

Castelli examined this issue in detail in the letter to Galileo of 15 August 1637, describing his experiments and explaining why the white and black sides of the brick absorbed heat differently. Castelli maintained that, in order to investigate this matter properly, it was necessary to attempt a definition of the notions of heat, light, and colour—a very difficult task, he wrote. To this end, he carried out a series of experiments in addition to developing geometrical demonstrations.[59] He aimed to determine whether the brick's differential heating was caused not only by light but also by fire, and whether the same effect occurred when the reverse side of the painted surface was exposed to light:

> I said that when I exposed the painted side of the brick to an ordinary fire of wood and left it for a little more than a quarter of an hour, I found that the heat had impressed itself almost equally in the black as in the white parts, that is, so little more in the black part that the difference was almost insensible. I also said that on exposing the reverse of the painted face

[56] Galilei 1968, 11:295. Favino 1997.

[57] Galilei 1968, 17:121, repr. in Castelli 1669, 49.

[58] Galilei 1968, 17:123 (this sentence is not in Castelli 1669). The translation of Castelli's letters is from Ariotti 1972.

[59] "E voglio dichiarare, in che cosa consista principalmente la mia difficultà con esempli geometrici. Io reputo assolutamente impossibile dimostrare una proprietà, o passione d'un suggetto del quale prima non sia stabilita, e supposta la sua definizione. [...] Forse potremo arrivare a qualche cognizione con supporre prima qualche notizia delle nature, e condizioni necessarie delle nominate cose, caldo, lume, nero, e bianco; e poi andarci avanzando a poco a poco al ricercato quesito [...]." (Castelli 1669, 65; 67). See also Castelli 1669, 62–63.

> of the brick to the light of the sun after the heat penetrated the thickness of the brick, the white part heated up as much as the black one. Finally I observed that on exposing the same painted face to the heat of a fire without light, the black and white parts heated up equally. These effects appear to me to be worth serious consideration, as we see in them a very marked difference among heat originating from fire without light, heat from light without fire, and heat originating in part from fire and in part from light. We see that the heat which proceeds from light only, heats up much more noticeably the black than the white (caeteris paribus). On the other hand, the heat from fire without light heats up equally the white and the black, while heat that originates from fire with light heats up a little more the black than the white.[60]

The experiments prompted Castelli to conclude that there exist different kinds of heat:

> From a careful observation of these data and from other findings in nature, we see a great difference in the operation of these heats. To give them names, we shall call those which are produced by light only luminous heat corpuscles (*calidi luminosi*) to differentiate them from those produced by fire without light which we shall call fiery heat corpuscles (*calidi fuocosi*). Then, we call mixed heat corpuscles (*calidi misti*) those which originate partly from light and partly from fire.[61]

Castelli's investigations were grounded in a corpuscular conception of light, in particular in the view that a luminous body emits particles that may be reflected or absorbed by the surfaces of bodies according to their colour. This is evidenced by the analogy he proposed between the sunbeam striking the brick and bullets fired at it. He reported asking his interlocutor—a young nobleman—the following question:

> Were we to shoot twenty-five pistol shots with red-hot balls into the black part [of the brick] and twenty-five into the white part and of those shot into the black part twenty were to bounce back but of those shot into the white part only five were to bounce back, in which of the two parts would there be more red-hot balls? In the black or in the white part?

The young man replied without hesitation that if this is the case, the white section would contain a greater number of balls. Castelli thus maintained that, if sunlight were distributed uniformly over the brick and the black portion nevertheless felt hotter than the white, this implied that a larger quantity of light corpuscles was retained by the black surface. Conversely, white reflects more than black and the white area contained fewer light corpuscles. As these corpuscles were held to be hot, the black surface became warmer than the white. At this point two major questions arose, questions which—Castelli stated—no scientist has thus far been able to resolve, namely "how light produces heat and why the white reflects and bounces back more light than the black."[62]

---

[60] Galilei 1968, 17:162–163; Castelli 1669, 67–68. Here the letter is incorrectly dated 15 August 1638, instead of 1637.

[61] Galilei 1968, 17:163; Castelli 1669, 68. Ariotti translated *calidi* with "heat corpuscles" etc. Though the translation gives the correct sense of Castell's argument, it is noteworthy that Castelli avoided speaking of atoms or of corpuscles. By contrast he employed the term "corpuscles" in the letter on the sunspots of 8 May 1612, see above. For the Catholic Church's censorship of atomism see Redondi 1985 and Donato 2003.

[62] Castelli 1669, 64.

He addressed these questions by investigating the difference between light and heat. Following in the footsteps of *The Assayer*, he maintained that light particles attain an extremely high velocity—perhaps even instantaneous—whereas heat corpuscles move in measurable time; moreover, their swiftness, subtlety, and penetrative power allow light corpuscles to diffuse across immense distances.[63] Owing to their extreme subtlety, light particles traverse the smallest interstices of bodies without meeting any obstacles and without colliding with other particles, however minute.[64] In order to explain the absorption of light by the black surface, Castelli appealed to its invisible texture. He conjectured that the particles of light were reflected by the microscopic layers composing the black surface and thus became, as it were, trapped within it.

> I therefore say that with some cogency we can define that quality which we commonly call blackness as being nothing more than a surface which acts as an artful tomb of light and which is so constructed that the corpuscles of light which strike it have all their segments of motion directed toward the interior parts of the body. Therein the corpuscles of light remain buried in the manner described.[65]

As they strike these 'planes,' the light particles excite their corpuscles and consequently generated heat:

> the light first entered those interstices however narrow. Then, it would strike the sides of these erect fibers and be reflected according to the inviolable rules of reflection, that is to say, at angles equal to those of incidence. The result would be that the light would seldom or never turn back toward its source. Thus, the light would become buried, so to speak, in the body, and the surface of this body would shine so very little to our eyes as to give that appearance which we call blackness. [...] we have highly probable grounds for saying that the corpuscles of light which move with the highest velocity, with fast and segmented movements can excite the particles of the above mentioned plane and hence, produce heat.[66]

By contrast, the texture of the white surface is so arranged that most corpuscles of light bounce back toward the exterior, so, they do not agitate the corpuscles of this part of the brick.[67]The difference between the effects of light and fire on the two colours of the surface lies in the mechanical properties of their respective corpuscles, smaller and swifter those of light, thicker and slower those of fire.[68] Finally Castelli

---

[63] "In oltre metto in considerazione un'altra grandissima differenza tra la luce, e il calore, la quale è che la velocità della luce è d'infinito intervallo superiore alla velocità del fuoco, come che quella arriva al sommo grado di velocità, e forse si fà in instanti, e questa si fà in tempo. [...] E per tanto da queste, e da altre condizioni, che si osserveranno in queste cose, inclino grandemente a pensare, che la luce sottilissima, velocissima, e penetrantissima operi, si sparga, e si diffonda per ispazi, e tratti immensi con esquisitissimi modi." (Castelli 1669, 68–69).

[64] Castelli 1669, 69.

[65] Castelli 1669, 72.

[66] Castelli 1669, 71, 73.

[67] Castelli 1669, 72–73.

[68] Castelli 1669, 73–74: "Di piu direi, per risolvere il dubbio per qual cagione esposto al fuoco il bianco, & il nero, allora il riscaldamento si fà quasi eguale e nel nero, e nel bianco, con quel poco, e quasi insensibile vantaggio di calore nel nero, della qual cosa penso, che la ragione sia, imperocche quanto alla parte del calore, che proviene dal fuoco, quel caldo fuocoso viene ad essere eguale nel

noticed that not only black and white heat up differently in sunlight—other colours do too. Colours that reflect more light heat up less, while colours that reflect less light heat up more.[69]

Galileo, then in house arrest at Arcetri, supported Castelli's investigation of light and heat.[70] It is noteworthy that Castelli, a Benedictine monk, refrained from using the term atom or corpuscles in his letters to Galilei of 1637 and formally recanted his previous adherence to atomism—a consequence of the Catholic Church censorship of atomism, which Redondi has well documented.[71] It is however apparent that Castelli's considerations on light and fire circulated among his friends and students, notably Giovanni Ciampoli, and members of the Accademia del Cimento.[72]

This very topic was investigated by Boyle too, who in *History of Colours* (1664) published his own account of the differential absorption of heat by colours. His experiment and the account he provided relied upon Castelli's to be found in the letters to Galilei of 1637. News of Castelli's research on heat and light might have reached Boyle via John Finch who kept Boyle informed of the experiments made in Tuscany.[73]

What follows is Boyle's experiment and his explanation, being almost identical with Castelli's:

> And to shew that the beams that fall on black bodies, as they do not rebound outwards to the eye, so they are reflected towards the body it self, as the nature of those erected particles, to which we have imputed blackness, requires, we will add an experiment that will also confirm our doctrine touching whiteness; namely, that we took a broad and large tile, and having whitened over one half of the superficies of it, and black'd the other, we expos'd it to the summer sun; and having let it lye there a convenient time (for the difference is more apparent, if it have not lain there too long) we found, as we expected, that whilst the whited part of the tile remained cool enough, the black'd part of the same tile was grown not only sensible, but very hot, (sometimes to a strong degree.) [...] of which the reason seems to be, that the superficial particles of black bodies, being, as we said, more erected, than those of white or red ones, the corpuscles of light falling on their sides, being for the most part reflected inwards from one particle to another, and thereby engag'd as it were and kept from rebounding upwards, they communicate their brisk motion, wherewith they were impell'd against the black body (upon whose account had they fallen upon a white body, they would have been reflected outwards) to the small parts of the black body, and thereby produce in those small parts such an agitation, as (when we feel it) we are wont to call heat.[74]

---

nero, e nel bianco, non essendo tanto sottile il caldo fuocoso, quanto il luminoso, & in conseguenza non potendo fare quegli scherzi, e giuochi cosi finiti, e regolati, come fà il caldo luminoso."

[69] Castelli 1669, 76.

[70] See Castelli's letter to Galilei of 26 September 1637, in Galilei 1968, 17:186.

[71] Redondi 1985, 560–561. Bucciantini 1992, 178–179.

[72] Favino 1997.

[73] Boyle 1999–2000, 4:40–41. For Finch, see Hutton 2009.

[74] Boyle 1999–2000, 4:74.

## 7.5 Conclusions

Francis Bacon's well-known kinetic theory of heat, articulated in the second book of the *Novum Organum*, was by no means an isolated position. It formed part of a broader efflorescence of corpuscular investigations into heat and fire. As we have seen, the reinterpretation of heat and fire as products of matter and motion was a prominent feature of the agenda pursued by early seventeenth-century advocates of atomism. Beeckman and Basson conducted experiments and formulated theories about the micro-mechanisms responsible for heat and fire. They appealed to the disposition of particles and pores—in other words, to the microstructure of bodies—to account for their visible properties. Basson, for his part, placed particular emphasis on ether, conceived as a very fine corporeal substance that filled the interstices between corpuscles and activated the motion of atoms.

Throughout his career, Galileo undertook experimental investigations of heat, fire, and light, interpreting these phenomena in corpuscular terms. In contrast to Bacon, however, Galileo declined to determine the essence of light and, more broadly, of the physical entities and processes he studied. An instance of this stance appears in his remarks on the invisible mechanism responsible for the explosive reaction of gunpowder, which he articulated in a letter to the eclectic Aristotelian philosopher Fortunio Liceti on 26 June 1640.

> And I shall consider myself most fortunate when, having understood what fire and light are, I shall be able to comprehend how a handful of cold, black gunpowder can contain twenty barrels of fire and many millions of lights; in addition to the fact that those tiny grains contain and hold fast, so to speak, a vast quantity of tiny bows which, when are shot, produce an astonishing force and speed. Here I would not want to be told that I should not be satisfied with the truth of the fact, since this is what experience shows me to be the case; which I could say that in all the effects of nature that I admire, it assures me of the *an sit* but it brings me no gain in the *quomodo*.[75]

In another letter to Liceti, dated 25 August 1640, he reaffirmed that, contrary to Giulio Cesare Lagalla's assertions, he had never explicitly endorsed the thesis that light was material and corporeal (*materiale e corporea*) in nature.[76]

By contrast, Castelli upheld a corpuscular theory of heat and light and offered a mechanical explanation of the relationship between heat, light, and colour. His account, although—as we have seen—circulating in manuscript from 1637, was published only in 1669. It is very likely Boyle was acquainted with this experiment (and with the accompanying mechanical interpretation). Boyle repeated it and further developed Castelli's investigations of heat and light by formulating a series of hypotheses on the invisible mechanical properties of corpuscles producing the coloured surfaces of bodies. Castelli's work on light and heat proved to be seminal, and it was further developed by early modern scientists such as Boyle, Newton, Boerhaave, and Franklin.[77]

---

[75] Galilei 1968, 18:208. See also 217.

[76] Galilei 1968, 18:233–234.

[77] Cohen 1955.

**Acknowledgements** The research presented in this paper was carried out with the financial support from PRIN 2022 PNRR Missione 4—Componente 2—Next Generation EU CODICE CUP: F53D23012640006. I would like to thank Michele Camerota who has provided useful comments on an earlier version of the manuscript.

## References

Ariotti, Piero. 1972. Benedetto Castelli: Early Systematic Experiments and Theory of the Differential Absorption of Heat by Colors. *Isis* 63 (1): 79–87.

Barry, Jonathan, and Fabrizio Bigotti, eds. 2022. *Santorio Santori and the Emergence of Quantified Medicine, 1614–1790*. Cham: Palgrave Macmillan.

Basson, Sébastien. 1621. *Philosophiae naturalis adversus Aristotelem libri XII*. Geneva: Apud Petrum de La Rovière.

Beeckman, Isaac. 1939–1953. *Journal tenu par Isaac Beeckman de 1604 à 1634*. Ed. Cornelis de Waard. 4 vols. La Haye: Martinus Nijhoff.

Bigotti, Fabrizio. 2017. A Previously Unknown Path to Corpuscularism in the Seventeenth Century: Santorio's Marginalia to the *Commentaria* in *Primam Fen Primi Libri Canonis Avicennae* (1625). *Ambix* 64 (1): 1–14.

Bigotti, Fabrizio. 2022. 'Gears of an Inner Clock': Santorio's Theory of Matter and Its Applications. In *Santorio Santori and the Emergence of Quantified Medicine, 1614–1790*, edited by Jonathan Barry and Fabrizio Bigotti, 65–101. Cham: Palgrave Macmillan.

Boyle, Robert. 1999–2000. *The Works of Robert Boyle*. Ed. Michael Hunter and Edward B. Davis. 14 vols. London: Pickering & Chatto.

Bucciantini, Massimo. 1992. Atomi, geometria e teologia nella filosofia galileiana di Benedetto Castelli. In *Geometria e atomismo nella scuola galileiana*, edited by Massimo Bucciantini and Maurizio Tonini, 171–191. Florence: Leo S. Olschki.

Camerota, Michele. 2025. *Galileo. Alle origini della scienza moderna*. Milan: Hoepli.

Cardano, Girolamo. 1550. *De subtilitate libri XXI*. Nuremberg: Johann Petreius.

Castelli, Benedetto. 1669. *Alcuni opuscoli filosofici…* Bologna: per Giacomo Monti, ad istanza degli eredi del Dozza.

Clericuzio, Antonio. 2000. *Elements, principles and corpuscles: A study of atomism and chemistry in the seventeenth century*. Dordrecht: Kluwer.

Clericuzio, Antonio. 2023. "The Best Philosopher in France" The Reception of Gassendi's Natural Philosophy in England. In *Pierre Gassendi. Humanism, Science, and the Birth of Modern Philosophy*, edited by Delphine Bellis, Daniel Garber, and Carla Rita Palmerino, 334–376. New York: Routledge.

Cohen, I. Bernard. 1955. Franklin, Boerhaave, Newton, Boyle & the Absorption of Heat in Relation to Color. *Isis* 46 (2): 99–104.

Donato, Maria Pia. 2003. L'onere della prova. Il Sant'Uffizio, l'atomismo e i medici romani. *Nuncius* 18 (1): 69–87.

Everson, Stephen. 1997. *Aristotle on Perception*. Oxford: Clarendon Press.

Favaro, Antonio. 1983. *Amici e corrispondenti di Galileo Galilei*. Vol. 3 vols. Florence: Salimbeni.

Favino, Federica. 1997. A proposito dell'atomismo di Galileo: da una lettera di Tommaso Campanella ad uno scritto di Giovanni Ciampoli. *Bruniana & Campanelliana* 3 (2): 265–282.

De Ferrari, Augusto. 1978. Benedetto Castelli. In *Dizionario Biografico degli Italiani*, vol. 21. Rome: Istituto della Enciclopedia Italiana.

Furley, David J. 1993. Democritus and Epicurus on Sensible Qualities. In *Passions and Perceptions: Studies in Hellenistic Philosophy of Mind*, edited by Jacques Brunschwig and Martha C. Nussbaum, 72–94. Cambridge: Cambridge University Press.

Galen. 1996. *Galeni De elementis ex Hippocratis sententia/Galen, on the elements according to Hippocrates*. Edited and translated by Phillip De Lacy. Berlin: Akademie Verlag.
Galilei, Galileo. 1957. *Discoveries and Opinions of Galileo*. Translated by Stillman Drake. Garden City, NY: Doubleday Anchor Books.
Galilei, Galileo. 1968. *Le opere di Galileo Galilei. Edizione nazionale a cura di Antonio Favaro et al*. Vol. 20 vols. Florence: Barbèra.
Galilei, Galileo. 2023. *Il Saggiatore, Edizione commentata a cura di Michele Camerota e Franco Giudice*. Milan: Hoepli.
Galluzzi, Paolo. 2011. *Tra Atomi e indivisibili. La materia ambigua di Galileo*. Florence: Leo S. Olschki.
Gassendi, Pierre. 1649. *Animadversiones in decimum librum Diogenis Laertii*. Lyon, Guillaume Barbier.
Gemelli, Benedino. 2002. *Isaac Beeckman. Atomista e lettore critico di Lucrezio*. Firenze: Leo S. Olschki.
Gómez, Susana. 2013. From a Metaphysical to a Scientific Object: Mechanizing Light in Galilean Science. In *The Mechanization of Natural Philosophy*, edited by Daniel Garber and Sophie Roux, 191–215. Dordrecht: Springer.
Hirai, Hiro. 2019. Telesio, Aristotle, and Hippocrates on Cosmic Heat. In *Bernardino Telesio and the Natural Sciences in the Renaissance*, edited by Pietro Daniel Omodeo, 51–65. Leiden: Brill.
Hutton, Sarah. 2009. Finch, Sir John (1626–1682), in *Oxford Dictionary of National Biography*. https://www.oxforddnb.com/view/10.1093/ref:odnb/9780198614128.001.0001/odnb-9780198614128-e-9439.Accessed 20 January 2026
Ingegno, Alfonso. 1980. *Saggio sulla filosofia di Cardano*. Florence: La Nuova Italia.
Kahn, Charles H. 1960. *Anaximander and the Origins of Greek Cosmology*. New York: Columbia University Press.
Knuuttila, Simo. 2008. Aristotle's Theory of Perception and Medieval Aristotelianism. In *Theories of Perception in Medieval and Early Modern Philosophy*, edited by Simo Knuuttila and Pekka Kärkkäinen, 1–22. Dordrecht: Springer.
Lüthy, Christoph H. 1997. Thoughts and Circumstances of Sébastien Basson. Analysis, Micro-History, Questions. *Early Science and Medicine* 2 (1): 1–73.
Mendelsohn, Everett. 1964. *Heat and Life. The development of the Theory of Animal Heat*. Cambridge, MA: Harvard University Press.
Moreau, Elisabeth. 2018. *Eléments, atomes & physiologie: le contexte médical des théories de la matière (1567–1634)*. PhD diss. Nijmegen: Université Libre de Bruxelles/Radboud Universiteit.
Owens, Joseph. 1982. Aristotle on Common Sensibles and Incidental Perception. *Phoenix* 36 (3): 215–236.
Palmerino, Carla Rita. 2011. The Isomorphism of Space, Time and Matter in Seventeenth-century Natural Philosophy. *Early Science and Medicine* 16 (4): 296–330.
Pasnau, Robert. 2007. Democritus and Secondary Qualities. *Archiv für Geschichte der Philosophie* 89 (2): 99–121.
Redondi, Pietro. 1985. Atomi, Indivisibili e Dogma. *Quaderni Storici* 20 (59/2): 529–573.
Redondi, Pietro. 1987. *Galileo: Heretic*. Princeton: Princeton University Press.
Salvia, Stefano. 2025. 'Atoms of Fire': Galileo's Unachieved Theory of Heat and the Beginnings of Thermometry (c.1603–1638). In *Early Modern Fire: Science, Technology, and the Urban Space*, edited by Gianenrico Bernasconi and Marco Storni, 49–89. Leiden: Brill.
Sarpi, Paolo. 1996. In *Pensieri naturali, metafisici e matematici*, edited by Luisa Cozzi and Libero Sosio. Milan: Ricciardi.
Shea, William. 1970. Galileo's Atomic Hypothesis. *Ambix* 17 (1): 13–27.
Suárez, Francisco. 1978–1981. In *Commentaria una cum quaestionibus in libros Aristotelis De anima*, edited by Salvador Castellote. Madrid: Sociedad de Estudios y Publicaciones.
Taape, Tillmann. 2025. Subtle Fire: Distillation as Low-Heat Technology and the Agency of Human Art. *Ambix* 72:243–267.

Taylor, C. C. W. 1999. *The Atomists: Leucippus and Democritus. Fragments*. Toronto: University of Toronto Press.
van Berkel, Klaas. 2013. *Isaac Beeckman on Matter and Motion. Mechanical Philosophy in the Making*. Baltimore: Johns Hopkins University.
van Raalte, Marlein. 2010. The Nature of Fire and Its Complications: Theophrastus' *De igne* 1–10. *Bulletin of the Institute of Classical Studies* 53 (1): 47–97.
Zanier, Giancarlo. 1977. Il macrocosmo corpuscolaristico di Sebastiano Basson. In *Ricerche sull'atomismo del Seicento*, 79–118. Florence: La Nuova Italia.

**Antonio Clericuzio** is Professor of History of Science at Roma Tre University. He has published extensively on Boyle, on early modern theories of matter and chemistry. He is currently working on a book on chemistry and medicine in Italy.

# Chapter 8
# Looking for an Authority: The Failed Revival of Anaxagoras's Fiery Aether in the Early Modern Period

**Eva Del Soldato**

**Abstract** A few lines in *On the Heavens* I.3 reference Anaxagoras's belief that celestial aether was actually fire, based on an etymology opposed to the one adopted by Aristotle. Although Aristotle quickly dismissed Anaxagoras's position, he later mentioned it again in *Meteorology* I.1. How was the text read in the sixteenth and seventeenth centuries, when the Aristotelian fifth essence ended up in the firing line amid a season of new observations and theological concerns? This article examines the early modern reception of Anaxagoras's doctrine of the fiery aether, considering Aristotelian commentaries as well as works by authors who sought to challenge Peripatetic cosmology for philosophical, scientific, and religious reasons. It particularly focuses on Matthias Frkić's *De caelesti substantia*, a 1646 treatise that aimed to promote Anaxagoras as an anti-Aristotelian authority aligned with Christian needs. By evaluating the limited impact of Frkić's work, this contribution seeks to explain why Anaxagoras never experienced a significant revival in early modern cosmological debates despite certain advantages of his tenets.

**Keywords** Anaxagoras · Aristotle · Early modern cosmology · Aether · Fire

Anaxagoras of Clazomenes (ca. 499 BC–428 BC) is not the first ancient thinker one would associate with the philosophical and scientific debates of the early modern period. Unlike Empedocles and Parmenides, Heraclitus and Democritus, he was not even included in the pioneering first printed edition of ancient philosophers' fragments, published in 1573 by Stephanus.[1] The central tenets of his thought were nonetheless accessible from other sources: Diogenes Laertius, the pseudo-Plutarchean *Placita philosophorum*, Lucretius, Cicero, Galen, Plato, but mainly from Aristotle and his commentators. Yet, these testimonies presented early modern

[1] Stephanus 1573. See Grafton 1988, 767; Primavesi 2011; Kraye 2021, 103.

E. Del Soldato (✉)
University of Pennsylvania, Philadelphia, USA
e-mail: evadel@sas.upenn.edu

 
E. Jung et al. (eds.), *The Forgotten Giants: Overlooked Revolutions in Natural Philosophy*, Studies in History and Philosophy of Science 70,
https://doi.org/10.1007/978-3-032-34008-5_8

readers with a fragmented and even contradictory image of Anaxagoras, and this ambiguity was reflected in the ways his thought was occasionally appropriated.[2]

On the one hand, Anaxagoras was celebrated as a link in the tradition of the pious *perennis philosophia*: Agostino Steuco (1497 or 1498–1548), for example, claimed that Anaxagoras held, because of the *nous*, a doctrine of creation close to Mosaic and Hermetic views.[3] Nicholas Cusanus (1401–1464) saw in the Anaxagorean principle "omnia in omnibus" a prefiguration of a mediated divine immanent omnipresence (and a century later Giordano Bruno (1548–1600) would radicalise this intuition).[4] Nevertheless, Anaxagoras's statement about the hand—that men are the most intelligent animals because they have hands, and not vice versa—mentioned by Aristotle in *On the Parts of Animals* (687a 7), exposed him as a denier of teleology: again Bruno recovered this motif to support the lack of difference between corporeal and incorporeal matter.[5] After all, allegations of atheism had always been associated with Anaxagoras's name, especially because of Diogenes Laertius's testimony (II.3) about a trial from which only Pericles's intervention ultimately saved the daring philosopher. Anaxagoras's alleged heresy was still debated in eighteenth-century erudite dissertations, including one *Quo jure antiqui quidam philosophi athei vocentur* printed in Ulm in 1709: there Anaxagoras was described as "infamis" because of his crime, even if at the end the author—Elias Bücklen—scolded the Athenians for having built a trial against the philosopher because of political reasons (hatred toward his friend Pericles) and pure superstition (they believed the Sun to be a divinity, while Anaxagoras described it as a burning stone). Indeed, the accusations were unfair according to Bücklen, who mentioned the divine *nous* as decisive proof of Anaxagoras's innocence.[6] Anaxagoras was equally absolved ("absolvendus!") of atheism by Jacob Friedrich Reimann in his *Historia universalis atheismi et atheorum falso et merito suspectorum* (1725).[7] Anaxagoras's name also resurfaced in debates on atomism, starting with Jacob Schegk (1511–1589), and continuing with Joseph Du Chesne (1544–1609) and Sébastien Basson (d. after 1625).[8] Although appreciating aspects of his philosophy, Ralph Cudworth (1617–1688) regarded Anaxagoras as unable to entirely abandon the views of the Ionian materialists and, therefore, embrace pious atomism.[9]

This contribution focuses on a specific episode in Anaxagoras's early modern reception, related to debates over the corruptibility of the heavens, with particular reference to his belief—reported by Aristotle—that the aether, the fifth element, was fiery. Again, when thinking today of ancient authorities that were used in the sixteenth

[2] Doxographical tradition on Anaxagoras was already available in the Middle Ages, see Bray 2021 and 2024.

[3] Steuco 1540, 9–12.

[4] Cusanus 1488, I: 3–72 and II: 371–407; Bruno 1583 and 1584.

[5] Bruno 1585.

[6] Bücklen and Herttenstein 1709, 12–13.

[7] Reimann 1725, 146–147.

[8] Hirai 2024.

[9] Cudworth 1678, 16, 112–113, 117, 124–125. On Cudworth see Levitin 2015, esp. 365–366.

and seventeenth centuries to undermine the Aristotelian cosmology—which divided the world into an incorruptible supralunary region made of aether and a sublunary one, subject to the interplay of the four elements (including fire)—one would think of Plato, or the Stoics with their pneuma, made of fire and air, or even of the Holy Fathers.[10] Yet, a passage in *On the Heavens* (and a similar one at the beginning of *Meteorology*) gained Anaxagoras some attention in this context. In this article, I will trace the discussion of Anaxagoras's position in several early modern commentaries to *On the Heavens*, highlight his occasional triumph, and suggest possible explanations for why, in the end, the ancient philosopher remained a secondary presence in these debates.

In the first book of *On the Heavens*, Aristotle argued for the eternity and incorruptibility of the heavens, claiming that their constitutive element, in turn, was eternal and incorruptible because devoid of a contrary: the aether. Closing this discussion (270b4–25), he offered three "empiric" confirmations for his position:

> [ ...] for all men have some conception of the nature of the gods, and all who believe in the existence of gods at all, whether barbarian or Greek, agree in allotting the highest place to the deity, surely because they suppose that immortal is linked with immortal [...]; so far as our inherited records reach, no change appears to have taken place either in the whole scheme of the outermost heaven or in any of its proper parts [...]; the name, too, of that body seems to have been handed down right to our own day from our distant ancestors who conceived of it in the fashion which we have been expressing. The same ideas, one must believe, recur in men's minds not once or twice but again and again. And so, implying that the primary body is something else beyond earth, fire, air, and water, they gave the highest place the name of *aether*, derived from the fact that it 'runs always' [*aei thein*] for an eternity of time. Anaxagoras, however, misuses this name, taking aether as equivalent to fire [*aithein*, that is "to burn".[11]

This latter passage can be coupled with one from the *Meteorology* (339b 19–27), in which Aristotle was more restrained in reporting Anaxagoras's position, presenting it as the result of his own interpretation ("seems to me") and saving part of its foundational reasoning:

> [...] it appears to be an old belief and one which men have held in the past, for the word "aether" has long been used to denote that [supralunary] element. Anaxagoras, it is true, seems to me to think that the word means the same as fire. For he thought that the upper regions were full of fire, and that men referred to those regions when they spoke of aether. In the latter point he was right; for men seem to have assumed that a body that was eternally in motion was also divine in nature; and, as such a body was different from any of the terrestrial elements, they determined to call it 'aether.'[12]

In both passages, Aristotle was not simply rejecting Anaxagoras's etymology—based on *aithein*, "to burn"—but the cosmology that was attached to it, according to which

---

[10] Grant 1994, 190 and 267; Barker 1991; Van Nouhuys 1998, 98–104.

[11] Translation by J. L. Stocks in Aristotle 1984, 450–451.

[12] Translation by E. W. Webster in Aristotle 1984, 556. Alexander of Aphrodisias 1545, 2v, emphasised that Aristotle was distinguishing between the ancients—who believed the heavens non-generable and incorruptible—and Anaxagoras who identified aether with fire. Anaxagoras was right, according to Aristotle, in attributing a divine character to the heavens, but not in associating it with fire.

a corruptible element—fire—would have found citizenship in the supralunary world. In the *Meteorology*, he invoked the first argument from *On the Heavens*, the divinity of the celestial regions, to find something salvageable in Anaxagoras's words.

Modern scholarship has challenged Aristotle's statement in *On the Heavens*.[13] In fact, Anaxagoras's aether should be understood in light of the symmetry that characterises his universe, and therefore not as fire, but as dry and refined air, (in contrast to the damp terrestrial air), from which fire develops. Philologists also agree that Anaxagoras's etymology was correct, contrary to the stance endorsed by Aristotle and, before him, by Plato in his *Cratylus* (410b), where aether is equated with "perennial running," *aei thein*.[14] Nonetheless, Aristotle's assertion that the ancients commonly used the term "aether" to describe the constitutive divine element of the heavens—different from the sublunar ones like fire—provided support to his otherwise seemingly unprecedented fifth element; and yet, already ancient interpreters would need to justify the Philosopher's claim and reconcile the alleged antiquity of the concept of incorruptible aether with a doctrine that appeared invented by Aristotle himself.[15]

In the early modern period, the lines in *On the Heavens* immediately preceding the section on Anaxagoras, which refer to the absence of records of changes that took place in the supralunary world as proof of its incorruptibility, were differently exploited and targeted with gusto by anti-Aristotelians like Francesco Patrizi (1529–1597) and proponents of a new cosmology, including Galileo Galilei (1564–1642).[16] But what about that sentence on Anaxagoras? Was it exploited against Aristotle? And did it affect Anaxagoras's reputation, granting him a role in the pantheon of early modern cosmology?

First, it must be noted that for many decades—even centuries—interpreters were relatively compact in reading the passage from *On the Heavens* without questioning Aristotle's accusations against Anaxagoras, often repeating what was stated by ancient commentators like Simplicius (VI c.). Simplicius wrote his commentary on the text with a target in mind, his contemporary Philoponus, who wanted to show that the "cosmos is perishable and comes to be at some time."[17] Paradoxically, Simplicius's commentary had the merit of saving substantial passages from Philoponus's anti-Aristotelian work, which survived in a fragmentary form.[18] When discussing 270b4–25, Simplicius focused on etymology and, using a principle of economy, noted that "Aristotle censures Anaxagoras for incorrectly deriving the word *aither* from *aithein*, that is, 'burn,' and consequently applying it to fire. For if this were the natural understanding of the word, we would also have called fire *aither*. Now what

[13] Lanza 1966, 67, 131–132, 192–193; Schofield 1980, 71; Kingsley 1995, 28–29.

[14] Sedley 1998, 143.

[15] Palmer 2000, 194–195.

[16] Patrizi 1581, 435–436; Galileo 1890–1909, 7:136; Del Soldato 2020, 127–137.

[17] Simplicius 2011, 60.

[18] Philoponus 1987.

is the point of using two words with one meaning and leaving what is meant by one of them in the dark?"[19]

Simplicius is often mentioned in early modern commentaries in the context of the Anaxagoras passage, but less often than Averroes (Ibn Rushd, 1126–1198). In his commentary *On the Heavens*, Averroes had provided an effective template to defuse Anaxagoras's position by reconstructing the false syllogism the ancient thinker would have allegedly used to reach his conclusions and assimilate fire and aether: aether is a light body that moves fast; fire is a light body that moves fast; therefore, fire is aether.[20] Averroes admitted that Aristotle was the first among the Greeks to endorse the incorruptibility of the aether and the world's eternity, unlike his fellow Greeks, who wanted the "coelum factum." When Aristotle mentioned the antiquity of his opinion, in an apparent contradiction to the novelty of his doctrine, Averroes concluded that he was undoubtedly referring not to other Greeks but to the Chaldeans.[21]

Averroes was very much present in Agostino Nifo's treatment of the passage. Nifo (1469/70–1538), in his commentary *On the Heavens*, paired Anaxagoras with Plato, who believed the heavens to be made of terrestrial elements, even if that would impose violently on them a circular movement (and "nullum violentum est perpetuum").[22] He noted that if Anaxagoras had wanted the heavens to be fiery *aequivoce*, that would have been acceptable, but since he endorsed the definition *univoce*, the position was untenable: the warmth of the fire would cause corruption and death. Nifo also reported the syllogism attributed by Averroes to Anaxagoras and declared that the eternity of the world was Aristotle's heresy and the reason why he had to flee Athens.[23] In 1570, Mainetto Mainetti (d. 1572) equally endorsed the Commentator in reading the passage and commending Aristotle's demonstrative

---

19 Simplicius 2011, 60.

20 Averroes 1550, 9v: "Et quia ipse fuit primo dicens ex Physicis Graecis quod caelum esset ingenerabile, dedit testimonium ex Chaldaeis, dicitur enim quod Philosophia ita perfecta erat tempore eorum, sicut tempore Aristotelis. Et dicit Et hoc videbant Antiqui et in eo convenerunt, etc., idest et illud quod relatum est ab Antiquis in natura istius corporis convenit ei, quod in eo diximus, scilicet quod est natura quinta, dicebant enim quod aliud a quatuor elementis et derivaverunt ei nomen a velocitate motus et eius permanentia, scilicet aethereum. Quapropter credimus quae illi dixerunt ipsum esse naturam quintam. Quod, cum sit natura quinta, non habet contrarium, si enim haberet contrarium, esset natura sexta. Contraria enim dicuntur esse paria, sicut in corporibus simplicibus quatuor elementorum, et quia non habet contrarium non habet materiam, neque est generabile, neque corruptibile. Et quia quidam Antiquorum peccaverunt in intellectu huius nominis, scilicet aether, ostendit Aristoteles peccatum suum dicendo Anaximander [*sic*] autem non utebatur, etc., id est quod non utebatur hoc nomine secundum veram interpretationem, cum existimabat quod hoc nomen simpliciter non dicitur, nisi de igne, et credidit propter hoc quod caelum esset ignis, dicit et est error in idiomate, scilicet quod hoc nomen non erit de igne, sed de velociter motis, et cum ignis sit levis, et velocis motus, existimavit quod hoc nomen diceretur de eo."

21 Cf. the *Bibliotheca historica* by Diodorus Siculus (II: 30). For a later use of the argument Piccolomini 1607, 63v and 98v.

22 Nifo 1549, 11v [*sed* 13v].

23 Nifo 1549, 16v: "Averroes ostendit syllogismum, quo Anaxagoras erravit. Syllogizavit enim sic aether est corpus leve, et velociter motum, ignis est corpus leve et velociter motum, igitur ignis est aether. Patet syllogismum esse in secunda figura ex duabus affirmativis."

methodology. Mainetti rebuked Galen, who had doubted that philosophers could ever reach certainty with their demonstrations about the necessary connection between the generability and corruptibility of the world.[24] In his 1565 commentary *On the Heavens,* Lucilio Filalteo (d. ca. 1578) presented an expanded etymological review to confute Anaxagoras. Still, he also took the issue seriously from a physical point of view and reacted *in via Aristotelis* against ten possible arguments in support of fiery heavens. Calling Anaxagoras "obstinatus et pertinax," Filalteo put the ancient's fiery aether in connection with the idea of the sun as a burning stone, the alleged cause of his trial.[25]

Other authors simply dismissed Anaxagoras: this was the case of Paolo Loredano (1539–1599).[26] Francesco Piccolomini (1523–1607), in his posthumously printed commentaries *On the Heavens*, praised the "ratio ex etymologia" offered by Aristotle in 270b 23–25, because it rests on the congruity between a name and the thing it indicates when wise men give names. Aristotle would use the word "aether"—by which the ancients named the heavens—to prove that the heavens were made of a substance different from the sublunar four elements and eternal. Anaxagoras believed aether was fire because he attributed a different origin to the word. Still, the first to reject Anaxagoras had not been Aristotle but Plato.[27] Despite some discrepancies in their use of the word "aether" (in *Timaeus* 58d, Plato did not identify "aether" with the heavens, but with the purest kind of air), both Plato and Aristotle agreed on its connection with an eternal movement. Yet Plato would have been the first to suggest it. For this reason, Aristotle would be a mere "instaurator" (someone who renews something pre-existing) of this opinion, but not its "inventor" (the one who came up with it).[28] Even if Piccolomini minimised Aristotle's merits to spotlight Plato, he was nonetheless critical of Anaxagoras.

The treatment of the corresponding passage in the *Meteorology* changes little to this picture: commentaries on the text focused on the etymological issue, recalling the parallel *locus* in *On the Heavens.* The commentary on *Meteorology* by Francesco Vimercato (d. 1569) is exemplary in this sense: Vimercato referred to the passage in *On the Heavens* to better understand Aristotle's words in the text he was commenting on. He also provided an extensive etymological digression on the word "aether," noting that Cicero supported the same solution proposed by Anaxagoras.[29] This observation about Cicero was subsequently included and absorbed in commentaries

---

[24] Mainetti 1570, 102–103. See also Galen 2005, II: 576–577; Galen 2023, 98–99; Pietrobelli 2013; Chiaradonna 2017.

[25] Filalteo 1565, 8, 51–52, 93.

[26] Loredano 1598, 53.

[27] Piccolomini 1607, 33v: "Anaxagoras tamen censuit id nomen fuisse deductum ex verbo [...] ardeo, quia Caelum censuit ignem, et ignem putavit esse principium [...] quae sententia primo a Platone, mox a Aristotele reiecta est."

[28] Piccolomini 1607, 33v: "hac autem ratione Aristoteles huius opinionis se non inventorem, sed instauratorem manifestat."

[29] Vimercato 1565, 14.

*On the Heavens*, for example by Filalteo and the *Conimbricenses*, suggesting a sort of osmosis in interpreting the *loci paralleli* in the two texts.[30]

Only isolated voices—outside of the university commentary tradition—were raised against Aristotle[31] and in favour of Anaxagoras about the aether. Among the few dissidents was Gian Francesco Pico della Mirandola (1469–1533), who sought to expose the contradictions of pagan philosophy, especially those of Aristotle. Pico wrote that Aristotelians preached the doctrine of the fifth essence, believing the Philosopher demonstrated it. In fact, authorities such as Atticus, Ambrose, Augustine, and Francis of Meyronnes all showed that the existence of the incorruptible fifth essence was anything but proven by Aristotle. Meyronnes, in particular, supported the idea of a fiery heaven, along with many unnamed Platonic thinkers. Pico claimed that Aristotle had unjustly condemned Anaxagoras's etymology of aether since the connection with "burning" would be even more valid.[32]

Of course, Aristotle's treatment of Anaxagoras on this topic did not escape the avenger of pre-Socratic philosophy *par excellence*, Francesco Patrizi. According to Patrizi, Aristotle was driven by an insane desire for glory. For this reason, he would have slandered his predecessors in his writings, quite effectively erasing them from the history of philosophy and sometimes appropriating their ideas.[33] The case of Anaxagoras's fiery aether would be no exception. Patrizi first cited the lines from *On the Heavens* to point out that Aristotle admitted he said nothing new and wanted to show his agreement with the ancients.[34] Subsequently, in a different section of his work, Patrizi defended Anaxagoras's etymology, claiming that the passages on his fiery aether in the *Meteorology* and *On the Heavens* would be in reciprocal contradiction ("Attamen quod hic [*On the Heavens*, I.3] vituperat I Meteoro[logicorum] cap. 3 laudat").[35] In fact, Patrizi claimed that in the *Meteorology*, Aristotle seemed to accept Anaxagoras's identification between fire and aether. Yet, Patrizi cheated to make this point. He reported a partial quotation from the *Meteorology*, which appeared to support his accusations:

> "Is enim qui appellatur aether, antiquam accepit appellationem, quam Anaxagoras sane igni idem putasse mihi videtur significare, supera enim plena ignis esse ipseque eam, quae ibi

---

[30] Filalteo 1565, 8; Collegium Conimbricensis 1606, 77–78.

[31] Pletho attacked Aristotle's fifth element, but from an eminently Platonic perspective. Indeed, he even claimed that Anaxagoras was the philosopher Aristotle mostly admired because of their shared atheistic inclinations (despite introducing the *nous*, in the case of Anaxagoras, or the Gods, in the case of Aristotle). Cf. Lagarde 1973.

[32] Pico della Mirandola [1520], CLXXXIIv: "neque propterea iure ab eo damnatur Anaxagoras, qui pro igne aetherem dixerit, nam aetymologiae ratio verti potest [...] et uti videtur accomodatius."

[33] Patrizi 1581. See also Muccillo 1975.

[34] Patrizi 1581, 229: "Quo loco animadvertendum est ex sua ipsius praesenti assertione cuncta dogmata suae philosophiae alias fuisse celebria eumque cum antiquioribus esse voluisse concordem."

[35] Patrizi 1581, 229.

est potentiam, aetherem vocare putavit. Hoc quidem recte decernens", hic ergo laudat quod ibi vituperaverat.[36]

Not including Aristotle's specification that what Anaxagoras saw correctly was merely the divine nature of the heavens, but presenting the Stagirite as claiming that Anaxagoras was right in identifying fire and aether, Patrizi could make a case for such a gross internal contradiction in the *corpus Aristotelicum*. Patrizi branded the etymology argument as mere "sophistica apparentia."[37] Discussing the nature of the heavens, he also pointed out that Plato, the Pythagoreans, all the ancient Greeks, the Egyptians, and the Chaldeans agreed on the heavens's fiery nature. Even if he did not single out Anaxagoras in this list, Patrizi reminded his readers that Anaxagoras correctly predicted the fall of a large stone from the heavens, presenting him as an ally against the incorruptibility of the heavens.[38]

In the same years Patrizi was composing his *Discussiones*, new observations of *novae* and comets prompted a decisive season of debates concerning the fluidity and the corruptibility of the heavens, placing the immutable aether in the firing line.[39] The alternatives offered by ancient thinkers to Aristotle's fifth essence began, therefore, to populate more intensively even the pages of interpreters defending aether. These included a young Galileo, who, in his 1584 *Tractatio de caelo*, insisted that the idea of fiery heavens was common among ancient philosophers, recovering his doxography from Philoponus, the same source used by Jacopo Zabarella (1533–1589) in his *De rebus naturalibus* (originally printed in 1590) to discuss celestial heat.[40] Zabarella claimed that certain unnamed ancient philosophers believed the stars were fiery because they relied too much on their senses.[41]

Even when openly mentioned by name, the presence of Anaxagoras continued to be almost negligible in these debates. In a work entirely devoted to the fifth essence and addressed against Xenarchus and Philoponus, Cesare Cremonini (1550–1631) merely reported the etymological disagreement between Anaxagoras and Aristotle.[42] Not different was the case of Tommaso Giannini (1556–1638), who dedicated several chapters of his 1618 *De substantia caeli* to refuting those thinkers who believed the

---

[36] Patrizi 1581, 303–304 (the passage appears to be based on Moerbecke's translation): "'That which is called aether received its name in ancient times, which, it seems to me, indicates that Anaxagoras indeed thought it to be the same as fire. For he believed that the upper regions are full of fire, and he called the force that exists there 'aether.' In this respect he was certainly judging correctly.' Therefore, here he praises what he had criticized there."

[37] Patrizi 1581, 435: "Et eum ante omnes qui ab etymo aetheris coeli aeternitatem astruere est conatus. Age eia egregie Thersites: ecquidnam est aeternitatis coeli, haec: Scientifica demonstratio? aut dialectica probatio? aut rhetorica persuasio? aut sophistica apparentia?."

[38] Patrizi 1581, 433 and 436. See also Vasoli 2002 and 2006.

[39] See at least Donahue 1981; Grant 1996 and 2003; Bucciantini 1999. See also Bianchi 1984 and Dales 1982 and 1987 for the medieval precedents of these debates.

[40] Galilei 1890–1909, I: 56–58.

[41] Zabarella 1617, 556.

[42] Cremonini 1616, 164. See also Cremonini 1613, Muir 2007, and Omodeo 2019.

heavens to be fiery, whether elemental or luminous, briefly mentioning Anaxagoras.[43] Although he equally aligned with the traditional Aristotelian solution, the Jesuit Bartolomeo Amico (1562–1649) also presented a complex dispute on the topic in 1648, including perspectives from ancient thinkers, Holy Fathers, and moderns alike, mentioning Anaxagoras's alternative etymology.[44]

On the other hand, scientific debates regarding the mutability of the heavens also provided new ammunition for attacks against Aristotle's eternal fifth essence based on theological and biblical arguments.[45] Authors such as Luis de Molina (1535–1600), Jacob Salianus (1558–1640), and Cornelio a Lapide (1567–1637), for example, proposed that the heavens were made of water.[46] Christoph Scheiner (1573–1650), who had initially opposed Galileo's interpretations of the sunspots, equally embraced fluid heavens (and a fiery sun) to declare their corruptible nature, subject to divine creation and apocalypse.[47] Notably, Scheiner devoted extensive sections of his *magnum opus*, the *Rosa ursina*, to listing theological, biblical, patristic, and astronomical sources, as well as ancient philosophical authorities supporting watery or fiery heavens. Among those, Anaxagoras made a fleeting appearance.[48]

Yet, in 1646, a combative Franciscan from Krk, Matija Frkić (1583–1669), professor of theology in Padua, published in Venice a treatise entitled *De caelesti substantia et ejus ortu ac motu in sententia Anaxagorae, philosophi celeberrimi.* With this work, Anaxagoras, until then a supporting actor, gained his first leading role in the debates on celestial substance, being presented as an authority who could offer a cosmological model sensible to religious needs.[49] Except for Antonino Poppi, scholarship has substantially ignored the treatise, but it has several reasons for interest.[50]

Embracing a common practice to legitimise pagan authors, Frkić opened the volume with a short biography of Anaxagoras, depicting him—among other things—as a sort of Christian who met the prophet Jeremiah in Egypt and whose soul was likely saved.[51] In the first book in the treatise, Frkić focused on Anaxagoras's cosmological doctrines. He praised the advantages of Anaxagoras's natural philosophy in comparison to Aristotle's from different angles, he showed the agreement of other ancient authorities with Anaxagoras, and he presented physical proofs against the

[43] Giannini 1618, 40–41: "Hinc Anaxagoras nomen aetheris, quo caelum intelligitur, duxit *apo to aithein* quod est urere, putavitque aetherem vehementi conversione lapides e terra abripuisse, atque hos adurendo astra procreasse, et alii Platone authore in Cratyllo [*sic*] caelum hoc nomine appellatum esse voluerunt *aei thei peri aera reon* hoc est semper currit circa aerem fluens, ignem vero fluere circa aerem nemo non dat, qui in Aristotelis doctrina non sit versatus."

[44] Amico 1648, 130.

[45] See for a general context Martin 2014.

[46] Cardoso 1673, 133.

[47] Scheiner 1626–1630.

[48] Scheiner 1626–1630, 640.

[49] Frkić 1646.

[50] Poppi 1992.

[51] Frkić 1646, 2 and 13–16. On biographies of ancient philosophers as paratexts in early modern texts see Palmer 2017.

Aristotelian denial of the fiery heavens. In doing this, Frkić included a section on the supralunary nature of comets, in which he introduced several modern allies, including Christophorus Clavius.[52] To Giannini, who placed little importance on the new observations of comets and insisted on the divinity of the heavens, Frkić replied with an accusation of paganism.[53]

Indeed, in the first part of the section entitled *De coeli ortu* it immediately becomes clear that Frkić used Anaxagoras's fiery heavens and recent observations as a springboard to address theological issues not dissimilar from those discussed by Scheiner and many others. If the heavens consist of a terrestrial element and are no different from the sublunary world, it follows that they cannot be eternal, just as taught by the Scriptures. Frkić employed heavenly imperfections as his main proof and asserted that Aristotle—embracing the eternity of the world—was not "assertor [...] paternae traditionis," but rather its "primus disertor."[54] Similar things could be said of Giannini, who abandoned the Scriptures in the name of Aristotle and, in doing so, violated the prescriptions of the papal Bull *Apostolici regiminis*.[55] In the following section, Frkić continued made Cremonini his primary target. As previously mentioned, Averroes had mocked Anaxagoras for the purported weak syllogism behind his identification of aether and fire. Frkić returned this accusation to its source, exposing Cremonini's poor logic based on the Commentator's arguments.

Evidently, Frkić utilised Anaxagoras in his book apologetically and strategically. The aforementioned biography—more of a hagiography—in the treatise's opening section is crucial for highlighting Anaxagoras's religious viability. To this end, Frkić conveniently revised interpretations of Anaxagoras's doctrines that he had previously presented. Indeed, judging from the works he published a few years before *De caelesti substantia*, Frkić did not appear to regard Anaxagoras with particular sympathy. In 1639, for instance, he referred to the doctrine of homeomeries as "absurdissima."[56] In that same text, while controversially presenting textual evidence that Aristotle did not reject *creatio ex nihilo*, Frkić listed Anaxagoras among those who instead staunchly supported the axiom "ex nihilo nihil."[57] However, in the biography of Anaxagoras that opens *De caelesti substantia*, Frkić presented an entirely different account on that subject, relying on Steuco, who associated Anaxagoras with Moses and Hesiod.[58] These ancients all believed that matter (that is, chaos) existed before the world. Still, that first matter is made "ex nihilo" by the first intellect (the Anaxagorean *nous*), while the other existing things derive from the first matter and are resolved into that. Frkić claimed that only in this latter qualified realm, proper to natural philosophy,

[52] E.g., Frkić 1646, 69, see also 154.

[53] Frkić 1646, 89: "hoc argumentum [...] est philosophi pure pagani, non Christiani, et requirit detestationem, non dissolutionem."

[54] Frkić 1646, 149:

[55] Frkić 1646, 155.

[56] Frkić 1639, 5.

[57] Frkić 1639, 3.

[58] Steuco 1540, 9.

would Anaxagoras endorse "ex nihilo nihil." The metaphysical priority of the *nous* would crucially limit the applicability of the axiom.[59]

The programmatic choice of Anaxagoras as the authority of reference is further evidenced by Frkić's selection of authors such as Cremonini, Giannini, and also Zabarella as his main opponents. As noted earlier, these authors said little or nothing explicitly against Anaxagoras. Yet, they resolutely defended Aristotelian cosmology. For Frkić, Anaxagoras served as a convenient mask, useful for advocating a cosmology alternative to Aristotle's—one that could better meet the needs of Christianity while still being validated by an ancient authority.

This position implied that Aristotle's philosophy was either impious or not sufficiently pious, and in promoting it, Frkić likely had a specific target in mind. By "baptising" Anaxagoras based on those few lines in *On the Heavens*, comparing him to Moses, and asserting that his soul was likely saved, Frkić was rejecting the effort made just a year earlier by his arch-nemesis, Fortunio Liceti (1577–1657), to present Aristotle as a pious Jew and, thus, a monotheist.[60] Liceti aimed to preserve Aristotle's privilege, while Frkić sought to give Anaxagoras a halo to undermine Aristotelianism even more than the Philosopher himself. The Stagirite was indeed represented in relatively fair terms throughout Frkić's treatise, while the Franciscan was wholly opposed to his ancient and modern interpreters and their adulatory exegetical excesses—precisely what he criticised Liceti for in many other writings.[61]

While Liceti did not apparently comment on *De caelesti substantia*, another Paduan interlocutor of Frkić, the Philosophy professor Johannes Cottunius (1577–1658), seemed to react against it. In his commentaries *On the Heavens*, Cottunius initially limited his remarks to the etymological issue concerning aether and observed that, according to some interpreters, Aristotle was unfair in his accusation against Anaxagoras, since he equally used the word aether to speak about fire in the fourth book of the *Physics* (212b 20–22). Cottunius replied to this allegation by explaining that, in the *Physics*, Aristotle was not talking of the fifth element but of the fire "supra omnem aerem expansum."[62] Relying on the methodological order of the Aristotelian *corpus*, Cottunius also claimed that, when he composed the fourth book of the *Physics*, the Philosopher had not yet demonstrated the incorruptibility of the heavens (as he will do in *Physics* VIII and *On the Heavens* I), so he was free to embrace, as in other instances, the linguistic use of his time. After Aristotle proved the incorruptibility of the heavens, he consistently employed the term aether just for the supralunar world.[63] In closing the section, Cottunius expanded his attacks, rebuking those moderns who recovered the ideas of other ancient philosophers simply because

[59] Frkić 1646, 11: "Factum est igitur Chaos a primo intellectu de nihilo, caetera quae fiunt, ex prima materia fiunt, in primam materiam resolvuntur. Prius assertum est primae philosophiae, nam in philosophia naturali, illud est Anaxagorae dogma, ex nihilo nihil, nihil in nihil interit."

[60] Liceti 1645. See Del Soldato 2017 and 2025.

[61] Del Soldato 2025.

[62] Cottunius 1653, 86.

[63] Cottunius 1653, 86: "Plurima occurrunt exempla, in quibus Aristoteles agens de re quapiam ante determinatum eius locum, verba facit ex communi illorum temporum opinione."

of their unjust hate of Aristotle, invoking divine mercy against their obnoxious ambition. Among the examples he offered, including supporters of Pythagoras's heavens and Democritus's atoms, Cottunius singled out unnamed authors who "Anaxagorae dogma de substantia caeli defendunt idque fortasse cum quodam non debito livore in Aristotelem."[64] Chronological circumstances, the close interactions between the two professors in Padua, and especially Frkić's distinctive allegiance to Anaxagoras suggest that Cottunius's words were specifically targeting his colleague. If this is accurate, Cottunius's reaction would be among the few responses provoked by Frkić's work.

Although Frkić's passionate plea for Anaxagoras did not earn the ancient philosopher widespread attention—much less a halo—observations aligned with theological concerns led supporters of fiery heavens to grow rapidly. Listing Anaxagoras among several other ancient thinkers, another Paduan professor interested in the pre-Socratics, Claude Bérigard (1578–1663), reminded his readers that the ancients envisioned the sun as fiery and that the Holy Fathers upheld this conviction.[65] He mentioned that Ambrose harshly rebuked Aristotle, asserting that one cannot expect any truth from the Philosopher since he denied such an obvious truth, now confirmed by the telescope. Bérigard went on to show that the Holy Fathers believed not just the sun but the entire heavens to be made of fire, and that the souls of holy men were purified by it, as Origen claimed. Like Frkić, Bérigard suggested that only the constraining hold of Peripatetic philosophy drove thinkers away from the true nature of the heavens, resulting in doctrines alien to the original ideas of the pre-Socratics.

Less critical of Aristotle, Jesuit astronomers like Giovan Battista Riccioli (1598–1671) and Melchior Cornaeus (1598–1665) also embraced the concept of fiery heavens. However, similar to Bérigard, their engagement with Anaxagoras was minimal. In the *Almagestum novum* (1651), Riccioli mentioned Anaxagoras among many other ancients who believed the sun to be a burning stone and the heavens around the moon to be fiery. In his 1657 *Curriculum philosophiae Aristotelicae,* Cornaeus referenced only Heraclitus and Pythagoras, along with the Holy Fathers and Doctors, when concluding that the heavens "probabilius est esse ignem." He did not include any mention of Anaxagoras, even if that was possibly caused by his reliance on Albert's *De coelo.* While in I.4, the reference quoted by Cornaeus, Anaxagoras is associated with Empedocles to Heraclitus and the Pythagoreans, in I.1, Albert expressly attributed the confusion between aether and fire to the Pythagoreans and Anaximander. Anaximander may have been a paleographical error for Anaxagoras (cf. n. 20 above).[66]

---

64 Cottunius 1653, 87: "'They defend Anaxagoras' doctrine concerning the substance of the heavens, and perhaps they do so with a certain unwarranted hostility toward Aristotle."

65 Bérigard 1647, 114.

66 Riccioli 1651, 92, 186; Cornaeus 1657, 490. On Riccioli see Graney 2015, on Cornaeus see Blum 2012, 127–135 and Del Soldato 2020, 137. See also Albert 1971, 24: "Pythagorici quidem sicut Anaximander et sui sequaces non utuntur nomine aetheris secundum veritatem nominis et secundum proprietatem rei qua imponitur nomen. Dicunt enim illi, quod ignis est aether, et errant in nomine decepti per hoc quod videbant, quod corpus caeleste inflammat et ignit vicinum corpus."

It is well recognised that early modern thinkers were open to reviving ancient philosophies, often with an anti-Aristotelian agenda in mind. Bernardino Telesio and Parmenides, Pierre Gassendi and Epicurus, Ralph Cudworth and the Italic philosophers serve as prominent examples of how criticisms against an ancient tradition—Aristotelianism—could be formulated using alternative ancient traditions, demonstrating how old authorities could operate in an *avant-garde* mode.[67] Anaxagoras did not find early modern sponsors except Frkić, who, after all, extolled the ancient philosopher from a narrow perspective. Frkić's pugnacious endeavour to assimilate Anaxagoras to Christian tenets—a "necessary prelude" to any early modern revival—was equally unsuccessful not only because of its fanciful reliance on biographical legends but also for other "structural" reasons.[68]

*In primis*, the brief lines in *On the Heavens*, which form the basis of Frkić's treatise, were merely a second-hand excerpt of Anaxagoras's doctrines. Other ancient authors, such as Philoponus, provided a comprehensive array of arguments against the Aristotelian fifth essence and overall cosmology. Likewise, in many respects, Plato offered teachings that were akin to Anaxagoras's ideas (including the fiery heavens and the divine intelligence ordering pre-existing matter) for which a broader interpretative context was available and which had already been widely embraced in a Christian context, rendering the testimony of the philosopher from Clazomenes in many ways redundant.[69] In the face of this competition, the fragmentary evidence concerning Anaxagoras's cosmological doctrines made it difficult to adopt him as a reference to oppose the declining Aristotelian paradigm.

Moreover, even if Aristotle himself had targeted Anaxagoras in *On the Heavens*, and even if anti-Aristotelians were always eager to find fault in the Stagirite's own words, the focus on etymology surrounding Anaxagoras's fiery heavens watered down the passage's physical implications for many readers. Those lines could be regarded as yet another example of Aristotle's bad faith, as suggested by Patrizi, or—because of the absence of further context—at most be capitalised to add Anaxagoras's name to mere lists of ancient thinkers opposed to Aristotle's positions. All this helps to understand why Anaxagoras was not fully vindicated even when the fifth essence was bidding farewell. This is, therefore, the tale of a recovery that never occurred and of an authority that never was.

**Acknowledgements** This work has been supported by the Croatian Science Foundation under the project IP-2025-02-1552.

[67] Rogers, Sorell and Kraye 2010.

[68] Bayle 1734, I: 305 defused for good the legend about Anaxagoras's alleged trip to Egypt to meet Jeremiah. See Kraye 2019, 162.

[69] Even Frkić insisted on the assonances between Anaxagoras and Plato. See, e.g., Frkić 1646, 45, 53, 212.

## References

Albert the Great. 1971. In *De coelo et mundo*. Ed. Paul Hossfeld. Monasterii Westfalorum: Aschendorff.

Alexander of Aphrodisias. 1545. *In quatuor libros Meteorologicorum Aristotelis, Commentatio lucidissima. Alexandro Piccolomineo interprete*. Venice: apud Scotum.

Amico, Bartolomeo. 1648. *In Aristotelis Libros de caelo et mundo dilucida textus explicatio et disputationes*. Naples: apud Roncaliolum.

Aristotle. 1984. In *The Complete Works*. 2 vols. Ed. Jonathan Barnes. Princeton: Princeton University Press.

Averroes. 1550. *Aristotelis Stagiritae De caelo, De generatione et corruptione, Meteorologicorum, De plantis libri cum Averrois Cordubensis variis in eosdem commentariis*. Venice: apud Iuntas.

Barker, Peter. 1991. Stoic Contributions to Early Modern Science. In *Atoms, Pneuma, and Tranquillity: Epicurean and Stoic Themes in European Thought*, edited by Margaret J. Osler, 135–154. Cambridge: Cambridge University Press.

Bayle, Pierre. 1734[5]. *Dictionaire historique et critique*, 5 vols. Amsterdam: par la Compagnie des libraires.

Bérigard, Claude. 1647. *Circulus Pisanus de veteri et peripatetica philosophia in Aristotelis Libros de coelo*. Udine: Schiratti.

Bianchi, Luca. 1984. *L'errore di Aristotele. La polemica contro l'eternità del mondo nel XIII secolo*. Florence: La Nuova Italia.

Blum, Paul Richard. 2012. *Studies on Early Modern Aristotelianism*. Leiden: Brill.

Bray, Nadia. 2021. Anaxagoras in the Late Middle Ages. A Doxographical Study of Thomas of York's *Sapientiale*. In *Past and Future. Medieval Studies Today*, edited by Maarten J. F. M. Hoenen and Karsten Engel, 317–336. Turnhout: Brepols.

Bray, Nadia. 2024. From Anaxagoras to Albert the Great: The Latency of Forms and the Active Power of Matter in the Middle Ages. *Noctua* 11: 368–392.

Bruno, Giordano. 1583. *Sigillus sigillorum*. London: John Charlewood.

Bruno, Giordano. 1584. *De la causa, principio et uno*. London: John Charlewood.

Bruno, Giordano. 1585. *Cabala del cavallo pegaseo*. Paris: Antonio Baio [i.e., London: John Charlewood].

Bucciantini, Massimo. 1999. Teologia e nuova filosofia: Galileo, Federico Cesi, Giovambattista Agucchi e la discussione sulla fluidità e corruttibilità del cielo. In *Sciences et religions de Copernic à Galilée (1540–1610)*, 411–442. Rome: École française de Rome.

Bücklen, Elias, and Johann Friedrich Herttenstein. 1709. *Quo jure antiqui quidam philosophi athei vocentur*. Ulm: Gassenmayer.

Cardoso, Isaac. 1673. *Philosophia libera in septem libros distributa*. Venice: Bertanorum sumptibus.

Chiaradonna, Riccardo. 2017. Galen and the Disputes of the Philosophers: the Generation of Cosmos. *Medicina nei secoli: Journal of History of Medicine and Medical Humanities* 29: 861–878.

Collegium Conimbricensis. 1606. *Commentarii in quatuor libros De coelo, Meteorologicos & Parva Naturalia Aristotelis Stagiritae*. Venice: apud Vincentium et Amadinum.

Cornaeus, Melchior. 1657. *Curriculum philosophiae Peripateticae, ut hoc tempore in Scholis decurri solet*. Würzburg: Zinck.

Cottunius, Johannes. 1653. *Commentarii in quatuor libros Aristotelis de Caelo una cum quaestionibus*. Padua: apud Lucianum.

Cremonini, Cesare. 1613. *Disputatio de coelo in tres partes divisa*. Venice: apud Balionum.

Cremonini, Cesare. 1616. *Apologia dictorum Aristotelis de quinta cœli substantia aduersus Xenarcum, Ioannem Grammaticum, & alios*. Venice: apud Meiettum.

Cudworth, Ralph. 1678. *The True Intellectual System of the Universe*. London: Roynston.

Cusanus, Nicholas. 1488. *Opera omnia*. 2 vols. Strasbourg: Martin Flach.

Dales, Richard C. 1982. Discussions of the Eternity of the World During the First Half of the Twelfth Century. *Speculum* 57: 485–508.

Dales, Richard C. 1987. Early Latin Discussions of the Eternity of the World in the Thirteenth Century. *Traditio* 43: 171–197.
Del Soldato, Eva. 2017. Saving the Philosopher's Soul: The *De pietate Aristotelis* by Fortunio Liceti. *Journal of the History of Ideas* 78:531–547.
Del Soldato, Eva. 2020. *Early Modern Aristotle. On the Making and Unmaking of Authority*. Philadelphia: University of Pennsylvania Press.
Del Soldato, Eva. 2025. Matej Frkić and His Enemies. In *Renaissance Aristotelianism in South-East Europe. Scholarship, Metaphysics, and Interactions with Platonism*, edited by Pavel Gregorić and Martino Rossi Monti, 125–148. London: Bloomsbury.
Donahue, William H. 1981. *The Dissolution of the Celestial Spheres, 1595–1650*. New York: Arno Press.
Frkić, Matthaeus. 1639. *Vestigationes peripateticae*. Padua: ex typographia Frambotti.
Frkić, Matthaeus. 1646. *De caelesti substantia et ejus ortu, ac motu in sententia Anaxagoræ philosophi celeberrimi*. Venice: Salis.
Fyrigos, Antonis. 1998. Joannes Cottunius di Verria e il neoaristotelismo padovano. In *Renaissance. Readings of the corpus Aristotelicum*, edited by Marianne Pade, 225–240. Copenhagen: Museum Tusculanum Press.
Galen. 2023. *On Avoiding Distress and On My Own Opinions*. Eds. Ioannis Polemis and Sophia Xenophontos. Berlin: De Gruyter.
Galen. 2005. In *De placitis Hippocratis et Platonis*. Ed. Phillip de Lacy. Berlin: Akademie Verlag.
Galilei, Galileo. 1890–1909. *Opere*. 20 vols. Ed. Antonio Favaro. Florence: G. Barbera.
Giannini, Tommaso. 1618. *De substantia caeli et stellarum efficientia disputationes aristotelicae*. Venice: apud Meiettum.
Grafton, Anthony. 1988. The Availability of Ancient Works. In *The Cambridge History of Renaissance Philosophy*, edited by Charles B. Schmitt, Quentin Skinner, Eckhard Kessler, and Jill Kraye, 767–791. Cambridge: Cambridge University Press.
Graney, Christopher M. 2015. *Setting Aside All Authority: Giovanni Battista Riccioli and the Science Against Copernicus in the Age of Galileo*. South Bend: University of Notre Dame Press.
Grant, Edward. 1994. *Planets, Stars, and Orbs: The Medieval Cosmos, 1200–1687*. Cambridge: Cambridge University Press.
Grant, Edward. 2003. The Partial Transformation of Medieval Cosmology by Jesuits in the Sixteenth and Seventeenth Centuries. In *Jesuit Science and the Republic of Letters*, edited by Mordechai Feingold, 127–156. Cambridge, MA: MIT Press.
Hirai, Hiro. 2024. Jacob Schegk, Simplicius and "Anaxagoras the Atomist": Aristotelian Commentaries and Renaissance Corpuscularism. In *Hylomorphism into Pieces. Elements, Atoms, and Corpuscles in Natural Philosophy and Medicine, 1400–1600*, edited by Nicola Polloni and Sylvain Roudaut, 307–329. London: Palgrave Macmillan.
Kingsley, Peter. 1995. Notes on Air: Four Questions of Meaning in Empedocles and Anaxagoras. *The Classical Quarterly* 45: 26–29.
Kraye, Jill. 2019. Renaissance Humanism and the Transformations of Ancient Philosophy. In *Beyond Reception. Renaissance Humanism and the Transformation of Classical Antiquity*, edited by Patrick Baker, Johannes Helmrath, and Craig Kallendorf, 149–162. Berlin: De Gruyter.
Kraye, Jill. 2021. Reconfiguring the Boundary between Humanism and Philosophy. In *New Horizons for Early Modern European Scholarship*, edited by Ann Blair and Nicholas Popper, 95–110. Baltimore: Johns Hopkins University Press.
Lagarde, Bernadette. 1973. Le *De differentiis* de Pléthon d'après l'autographe de la Marcienne. *Byzantion* 43: 312–343.
Lanza, Diego. 1966. *Anassagora: testimonianze e frammenti*. Florence: La Nuova Italia.
Levitin, Dmitri. 2015. *Ancient Wisdom in the Age of the New Science: Histories of Philosophy in England, C. 1640–1700*. Cambridge: Cambridge University Press.
Liceti, Fortunio. 1645. *De pietate Aristotelis erga Deum et homines*. Udine: Schiratti.
Loredano, Paolo. 1598. *In Aristotelis De cœlo, libros quatuor commentaria*. Venice: apud Ciottum.

Mainetti, Mainetto. 1570. *Commentarii in lib.1. Aristotelis De coelo, necnon in librum Averrois De substantia orbis*. Bologna: Typis Rossii.
Martin, Craig. 2014. *Subverting Aristotle. Religion, History, and Philosophy in Early Modern Science*. Baltimore: Johns Hopkins University Press.
Muccillo, Maria. 1975. La storia della filosofia presocratica nelle *Discussiones Peripateticae* di Francesco Patrizi da Cherso. *La cultura* 13: 48–105.
Muir, Edward. 2007. *The Culture Wars of the Late Renaissance: Skeptics, Libertines, and Opera*. Cambridge, MA: Harvard University Press.
Nifo, Agostino. 1549. *Aristotelis Stagiritae De coelo & mundo libri quatuor*. Venice: apud Scotum.
Omodeo, Pietro Daniel. 2019. A Cosmos Without a Creator: Cesare Cremonini's Interpretation of Aristotle's Heaven. *Journal of Early Modern Studies* 8: 9–42.
Palmer, Ada. 2017. Humanist Lives of Classical Philosophers and the Idea of Renaissance Secularization: Virtue, Rhetoric, and the Orthodox Sources of Unbelief. *Renaissance Quarterly* 70: 935–976.
Palmer, John A. 2000. Aristotle on the Ancient Theologians. *Apeiron* 33:181–206.
Patrizi, Francesco. 1581. *Discussionum peripateticarum tomi quattuor quibus Aristotelicae Philosophiae universa Historia atque Dogmata cum Veterum Placitis collata, eleganter et erudite declarantur*. Basel: apud Pernam.
Philoponus, John. 1987. *Against Aristotle on the Eternity of the World*. Ed. Christian Wildberg. London: Duckworth.
Piccolomini, Francesco. 1607. *In Libros Aristotelis De coelo lucidissima expositio*. Venice: apud De Franciscis.
Pico, Giovan Francesco. 1520. *Examen vanitatis doctrinae gentium et veritatis Christianae disciplinae*. [Mirandola: Maciochius].
Pietrobelli, Antoine. 2013. Galien agnostique. Un texte caviardé par la tradition. *Revue des études grecques* 126:103–135.
Poppi, Antonino. 1992. Il *De caelesti substantia* di Matteo Ferchio tra tradizione e innovazione. In *Galileo e la cultura padovana*, edited by Giovanni Santinello, 13–56. Padua: CEDAM.
Primavesi, Oliver. 2011. Henri II Estienne über die philosophische Dichtung: eine Fragmentsammlung als Beitrag zu einer poetologischen Kontroverse. In *The Presocratics from the Latin Middle Ages to Hermann Diels. Akten der 9. Tagung der Karl und Gertrud Abel-Stiftung vom 5.–7. Oktober 2006 in München*, edited by Oliver Primavesi and Katharina Luchner, 157–196. Stuttgart: Franz Steiner.
Raphael, Renee. 2016. Eclecticism as a Vibrant Philosophical Program: Claude Berigard and Mauro Mancini on the University of Pisa. *History of Universities* 29: 1–24.
Reimann, Jakob Friedrich. 1725. *Historia universalis atheismi et atheorum falso et merito suspectorum*. Hildesheim: Schroeder.
Riccioli, Giovan Battista. 1651. *Almagestum novum astronomiam veterem novamque complectens observationibus aliorum et propriis*. Bologna: ex typographia heredis Benatii.
Rogers, G. A. J., Tom Sorell, and Jill Kraye, eds. 2010. *Insiders and Outsiders in Seventeenth-Century Philosophy*. London: Routledge.
Scheiner, Christoph. 1626–1630. *Rosa Ursina, sive, Sol ex admirando facularum & macularum suarum phoenomeno varius*. Bracciano: apud Andream Phaeum.
Schofield, Malcolm. 1980. *An Essay on Anaxagoras*. Cambridge: Cambridge University Press.
Sedley, David. 1998. The Etymologies in Plato's Cratylus. *The Journal of Hellenic Studies* 118: 140–154.
Simplicius. 2011. In *On Aristotle On the Heavens 1.3–4*, edited by Ian Mueller. London: Bloomsbury.
Stephanus, Henricus. 1573. Poiēsis philosophos. Poesis philosophica, vel saltem reliquiae poesis philosophicae Empedoclis, Parmenidis, Xenophanis, Cleanthis, Timonis, Epicharmi; adiuncta sunt Orphei illius carmina qui a suis appellatus fuit o theologos; item, Heracliti et Democriti loci quidam et eorum epistolae. Geneva: Stephanus.
Steuco, Agostino. 1540. *De perenni philosophia libri X*. Lyon: Gryphius.

Van Nouhuys, Tabitta. 1998. *The Ages of Two-faced Janus: The Comets of 1577 and 1618 and the Decline of the Aristotelian World View in the Netherlands.* Leiden: Brill.
Vasoli, Cesare. 2002. Sophismata putida: la critica patriziana alla dottrina peripatetica dell'eternità e immutabilità del cielo. In Francesco Patrizi filosofo platonico nel crepuscolo del Rinascimento, edited by Patrizia Castelli, 167–180. Florence: Olschki.
Vasoli, Cesare. 2006. Dissoluzione della "sfera eterna" e immutabile del cielo, nella polemica di Francesco Patrizi. In De Florence à Venise: études en l'honneur de Christian Bec, edited by François Livi and Carlo Ossola, 361–378. Paris: Presses de l'Université Paris-Sorbonne.
Vimercato, Francesco. 1565. *In quatuor libros Aristotelis Meteorologicorum commentarii.* Venice: Apud Scotum.
Zabarella, Iacopo. 1617. *De rebus naturalibus Libri XXX*. Frankfurt: Zetzner.

**Eva Del Soldato** is Associate Professor of Italian Studies at the University of Pennsylvania. Her research is primarily devoted to Renaissance thought and culture, with special attention to the Aristotelian and Platonic traditions. She is the author of the monographs *Simone Porzio* (2010) and *Early Modern Aristotle. On the Making and Unmaking of Authority* (2020). She has also published several articles and editions, including the Italian translation of Bessarion's *In calumniatorem Platonis* (2014). She has co-edited the volumes *City, Court, Academy. Language Choice in Early Modern Italy* (2017), *Harmony and Contrast. Plato and Aristotle in the Early Modern Period* (2022), and *Teaching Plato in Italian Renaissance Universities* (2024).

# Chapter 9
# Adam Chałupczyński (1822–1887): A Polish Pioneer in Evolutionary Thought

**Michał J. Wagner**

**Abstract** The historiography of evolutionary biology has long portrayed Charles Darwin as the solitary originator of the theory of evolution by natural selection, a view reinforced both by his own retrospective accounts and by the modern evolutionary synthesis. This article challenges that Darwin-centred narrative by examining the largely overlooked evolutionary theory of the Polish physician Adam Chałupczyński, developed independently in the 1850s and published in 1860. Chałupczyński advanced an account of species transformation grounded in the survival of the best adapted in the struggle for existence, drawing in part on Malthusian ideas but articulated within a distinct cultural and intellectual context. Rather than treating his theory as an incomplete or derivative precursor to Darwinism, the article situates it within nineteenth-century natural philosophy and analyses it as an original selectionist conception expressed through theological and cosmological language. By reconstructing Chałupczyński's views and their reception, the article argues that the idea of natural selection could emerge independently of Darwin and was not confined to British science. This case supports the broader claim that selectionism was, in a meaningful sense, "in the air" in the nineteenth century and could assume diverse conceptual forms.

**Keywords** Adam Chałupczyński · Natural selection · Critique of Darwinism · History of polish science · Nineteenth-century evolutionary thought

## 9.1 Introduction

In the historiography of biology, Charles Darwin has long been portrayed as the solitary creator of evolutionary theory—a view reinforced both by his own recollections and by later biologists and historians of the modern synthesis. In his *Autobiography*,

M. J. Wagner (✉)
Institute of Philosophy, Cardinal Stefan Wyszyński University in Warsaw, Warsaw, Poland
e-mail: m.wagner@uksw.edu.pl

 
E. Jung et al. (eds.), *The Forgotten Giants: Overlooked Revolutions in Natural Philosophy*, Studies in History and Philosophy of Science 70,
https://doi.org/10.1007/978-3-032-34008-5_9

Darwin recalled the attitudes of naturalists toward species variability around the time of *On the Origin of Species* (1958, 123–124):

> It has sometimes been said that the success of the *Origin* proved that the subject was in the air or that men's minds were prepared for it. I do not think that is strictly true, for I occasionally sounded a few naturalists and never happened to come across a single one who seemed to doubt the permanence of species.

This statement—part of Darwin's self-fashioning (or more precisely, of the image shaped by his son Francis)—fit neatly into the heroic narrative that cast him as the scientist who, through perseverance and sacrifice, formulated a theory that transformed natural history. In the twentieth century, this image was further entrenched by proponents of the neo-Darwinian synthesis (Browne 2010, 359–370). As the synthesis consolidated its dominance in evolutionary biology, its architects increasingly shaped the historiography of science as well. Darwin came to be established as the central—indeed, the sole—naturalist credited with producing a mature theory of evolution.[1] It might therefore appear that later historians often took his claim about the supposed absence of interest in evolution as an accurate reflection of the intellectual climate of pre-Darwinian biology. Yet this was not the case: evolutionary ideas, sometimes invoking mechanisms reminiscent of natural selection, had appeared repeatedly before 1859.

This article examines a theory that is usually overlooked in contemporary historiography: that of the Polish physician Adam Chałupczyński. He developed this theory from 1851 onwards, publishing it in 1860 in *Pomysły do wiedzy dziejów życia świata* ["Ideas Toward a Knowledge of the History of Life on Earth"]. Independently of Darwin, Chałupczyński proposed a theory of species evolution grounded in a process resembling natural selection. His work demonstrates that ideas akin to natural selection could emerge in cultural contexts far removed from Britain, and that they should not be viewed exclusively as products of the Victorian intellectual milieu. Moreover, contrary to Darwin's insistence, Chałupczyński's example suggests that the idea of evolution through selection was indeed "in the air" and could be formulated by more than one naturalist. The article is structured in four sections. The first two outlines Chałupczyński's biography and his evolutionary views in *Pomysły*.... The third discusses the reception of his work in Polish literature and explores possible sources of inspiration, clarifying the nature of his evolutionary views. The fourth and final section analyses his critique of Darwin's theory.

[1] This distinction is clearly reflected in Ernst Mayr's (1982, 352) division between those theories that can properly be called evolutionary—chief among them Darwin's theory of natural selection—and those that may be described only as so-called "transmutation theories," to which all of Darwin's precursors would have to be assigned.

## 9.2 Adam Chałupczyński (1822–1887)—A Biographical Sketch[2]

When examining the views of any Polish philosopher or naturalist of the nineteenth century, one must recall that Poland did not exist as a state: its territory was divided among Prussia, Austria, and Russia. Polish people preserved their national identity, even after the successive uprisings and independence movements failed. The history of this struggle is also the history of Adam Chałupczyński and is directly tied to the genesis of his major work, *Pomysły...* Chałupczyński was born in 1822 in Piotrków Trybunalski, in the Łódź district of the Kingdom of Poland, which was established in 1815. His childhood coincided with a brief period of relative political freedom: although the Kingdom was linked to Russia by a personal union, it possessed its own government and institutions such as the Royal University of Warsaw. This autonomy ended after the failed November Uprising of 1831: the Kingdom was fully incorporated into Russia, the University was dissolved, and General Ivan Paskevich became viceroy. The so-called "Paskevich era," culminating in martial law between 1833 and 1856, saw a drastic reduction in schools—only those sanctioned by a Government Commission created by Paskevich could remain. Many institutions were relocated from larger to smaller towns; the gymnasium from Kalisz was transferred to Piotrków, and this is where Chałupczyński studied (Kucharzewski 1914, 157–161). In 1837 he graduated from the Piotrków gymnasium, and a year later, at sixteen, began training as a pharmacist. After seven years he qualified as an assistant pharmacist, enabling him to enter the medical faculty at Moscow University. He completed his studies in 1849, earning a medical diploma. That autumn, the authorities appointed him physician of Kozienice in the Warsaw Governorate, bringing him back to his homeland. Within six months, however, he was arrested for involvement in the conspiracy of Szymon Konarski[3] and was imprisoned in the Zamość fortress from 1850 to 1853. After his release, he resumed medical practice. In 1859 he became a physician at the manor hospital in Maciejowice, serving as surgeon and obstetrician. During the January Uprising of 1863, insurgent units under Colonel Walenty Lewandowski were organised in Maciejowice. Chałupczyński joined them, though it is uncertain whether he commanded a detachment or served as Lewandowski's associate. He was sentenced to death by a military court for his role, but the sentence was not carried out. Instead, he aided the uprising as a physician, directing a field hospital in Izbica Kujawska. Chałupczyński remained in Izbica for thirteen years, from 1863 working as resident physician for the Miączkowski family. In 1876 he moved to Włocławek, where he lived until his death in 1887.

---

[2] All biographical information about Adam Chałupczyński is based on Dorcz 2016; Seyda 1999; Zawadzki n.d.

[3] Szymon Konarski (1808–1839) was a Polish revolutionary who was executed by the Russians in 1839 for his revolutionary activities. Chałupczyński was most likely accused of membership in the Związek Ludu Polskiego [Union of the Polish People], the organization Konarski had founded in 1837 with the aim, among other things, of preparing a new uprising (Mościcki 1949, 70–77).

## 9.3 The Theory of Evolution in *Pomysły Do Wiedzy dziejów życia*

As Chałupczyński himself admits in the preface to *Pomysły...*, he worked on the book for nine years, refining a conception that, in his view, offered a completely new perspective on the workings of nature. For this reason, he warned his readers in advance about the potentially controversial claims he intended to present (Chałupczyński 1860, I–II). As Bronisław Rejchman later observed (1869, 5), the work must in fact be regarded as unfinished—a point Chałupczyński himself acknowledged in the title. The book bore the name *Pomysły...* [*Ideas...*], because the author had not managed to assemble sufficient evidence to substantiate his theses fully. This circumstance can be explained by the conditions under which Chałupczyński worked. The preface is dated 15 April 1859—seven months before the publication of Darwin's *Origin...*—which suggests that he had begun writing as early as 1850, during his imprisonment in the fortress of Zamość. The *Pomysły...* were approved for publication by the Censorship Committee in December 1859, as evidenced by the censor Antoni Funkenstein's statement printed at the beginning of the volume. The book finally appeared in 1860, published at Chałupczyński's own expense. These details are crucial for any comparison between Chałupczyński's conception and Darwin's theory, for they show that the Polish physician arrived at similar conclusions independently of the English naturalist.

Chałupczyński's work is ambitious in scope: he seeks to outline the history of the world's development from the origins of the universe to the appearance of humankind. Yet the book remains only what its title promises—*Ideas...*—and does not aim to provide a complete explanation of nature's workings. Rather, it was intended as a starting point for further scientific inquiry:

> I remind the reader that I am only sketching ideas for a knowledge of the history of nature, not presenting in full those concepts which, in time, humanity will acquire about all things. If by this writing I can remove but one absurdity, dispel even a single prejudice that obstructs the path to truth, it will not be without benefit to my fellow men. (Chałupczyński 1860, 111)

How, then, did he intend to reach this goal? First, by changing the mode of scientific work. As he argued, natural science was too fragmented: numerous theories had been proposed, each describing a different phenomenon, yet without any unifying connection between them. From this multiplicity arose an epistemic chaos lacking proper synthesis. But since there is only one nature, all of its aspects must be interconnected, and there ought to exist a single theory encompassing the whole of phenomena (Chałupczyński 1860, IV, 135–136). Second, Chałupczyński called for a methodological shift in the natural sciences. He criticised extreme empiricism, maintaining that scientific inquiry need not consist exclusively of experimental data. Through our senses we perceive all that is necessary for understanding reality, and on this basis we can arrive at the fundamental laws governing its operation.

> In the study of all mysterious forces, whether spoken of in the past, present, or future, the surer path is always the intellectual one—but only such a path as does not depart from sensory observation, but proceeds alongside it. (Chałupczyński 1860, 82)

Thus, the requirement of presenting empirical evidence served chiefly to persuade others of one's claims; it was not, in itself, indispensable for attaining truth, which could also be reached through intuition (Chałupczyński 1860, 135). Empiricism, in his view, imposed unnecessary constraints in the form of rigid methodological rules, whereas most scientific problems could be solved if purely rational argument were admitted. In this context, Chałupczyński also polemicised with the prominent Polish naturalist Jędrzej Śniadecki, who had asserted that the origins of life were inaccessible to scientific investigation. For Chałupczyński, by contrast, such knowledge was indeed attainable, provided one allowed for the formulation of hypotheses concerning the first causes of life on the basis of observable phenomena (1860, 111–113).

Chałupczyński's methodology rests on the formulation of certain rationally acceptable fundamental hypotheses, from which further theses can be deduced and subsequently confirmed either by common experience or by experimental observation (cf. Chałupczyński 1860, 58–60, 241). He likely adopted this method from Lorenz Oken, whose philosophy of nature was also grounded in a deductive procedure (cf. Neuser 2001, 211–212). From Oken he also borrowed the notion that the universe is a "pulsating sphere" (Chałupczyński 1860, 1, 16), in which the relations among phenomena are determined by arithmetical principles. In this sense, all beings arise from varying proportions of primordial matter (1860, 132). This primordial matter consists of a finite number of atoms—called by Chałupczyński "niedziałki"—which remain in perpetual motion, combining and dissolving, thereby giving rise to new entities (1860, 26–27). All entities arise from the union of two different atoms, one passive and foundational, enabling the emergence of new qualities, and the other active, serving as the stimulus of transformation (1860, 8–9). This "creative force" of the atoms is one of Chałupczyński's fundamental, though indemonstrable, assumptions. The circulation of creation and destruction underlies both physical and biological transformations. Every being arises from primordial matter, then undergoes dissolution, so that from its remnants new forms may emerge (1860, 7–8). At the root of this process lies the oscillation radiating outward from the centre of the universe. At the point where the rays of the cosmic sphere converge, wave-like motion arises: matter presses against itself, producing the dynamics that drive the union and dissolution of atoms.

> These rays constitute the world-axes of nature; thus individual worlds seem to relate to one another as particles of the rays of a single immeasurable sphere. By means of these axes, like the strings of a cosmic orchestra, resounds the harmony of the world—once imagined—which only God himself can hear. From that centre the Creator strikes the rhythm of nature's ensemble and breathes his spirit into the bosom of humanity from time to time. Towards him, from the infinite, all bodies press upon each other, and that pressure arouses a wave-like motion along all the axes [...]. (Chałupczyński 1860, 16–17)

The movement of atoms and the manner of their combination depend on God's creative act. Chałupczyński explains the relation between God and nature in the following terms: "Nature, as an eternal and infinite whole, may rightly be called God's visible body and the sensible manifestation of his power." (1860, 1). The universe—understood as eternally pulsating matter—exists in an infinite, eternal, and unconditional mode. By contrast, whatever emerges within it, from the combination of atoms, is temporal and conditional. In this sense, the universe, as the foundation of reality and the repository of all that potentially exists within it, constitutes the physical representation of the divine. What is transient exists only momentarily in material form, yet in the thoughts of the Absolute it endures eternally as an idea. From the infinity of reality, one may therefore infer the existence of God—though ultimately this remains a matter of faith (1860, 2–4). This final point marks another of Chałupczyński's basic hypotheses about natural history. According to him, the oscillations of the universe, set in motion by God, drive the circulation of matter, which manifests itself in the gradual transformation of reality. This process is finalistic, and follows particular order: transformations of living world are aimed at the achieving perfection understood as likeness to God. This goal will be achieved when humanity attains the requisite level of spiritual development and becomes worthy of salvation (1860, 6, 46–47). Once this goal is realised, matter will disintegrate, and the process of creation will begin anew (1860, 47).

The foundation of Chałupczyński's evolutionism is thus the process of atomic circulation, which culminates in the advent of conditions enabling the salvation. The ontological diversity produced in this process arises not only from internal atomic combinations, but also from interactions with the environment—that is, with entities formed through other atomic configurations—which together determine the properties of individual beings. The final form of any entity therefore depends both on its constituent elements and on its environment (1860, 55, 58–59). In this way, the entire universe is organised into a network of interconnections, driven by the eternal pulsation of the cosmic "strings." In short, this means that Chałupczyński seeks to explain every phenomenon as the joint product of atomic composition and external influences. Thus, for example, he interprets light as a natural property of entities whose atoms vibrate in the proper rhythm, while simultaneously being influenced by the Earth's atmosphere and the Sun (1860, 61–67). In similar fashion he accounts for the phenomena of gravity (1860, 36–37), planetary orbital motion (1860, 37–39), electricity (1860, 88–102), and ultimately the origin of life.

The emergence of life depends on the formation of three fundamental kinds of terrestrial matter: "kruszców" [ore], ie.: solid, mineral bodies; "cieczy" [liquids], above all water; and "lotków" [volatiles] such as air and other gaseous substances. Every being on Earth is a combination of these three forms of matter (1860, 117, 129–130). For life to arise, conditions must exist in which these three elements interact under the influence of magnetic and electrical forces. Life therefore originates abiogenetically on seashores—places where minerals, air, and water converge. Chałupczyński suggests that this most likely occurs in sandy mud, with the first form of life appearing as a single-celled organism (1860, 165–167). Crucially, the transformation of inorganic into organic beings need not have been a one-time event: matter itself possesses the inherent potentiality to generate life. Everything depends solely on the presence of suitable conditions for the proper conjunction of the three kinds of matter (1860, 162–165). This marks the beginning of the evolutionary process, driven by the striving of organisms for self-preservation and for continual self-improvement. On the physicochemical level, this evolution results both from environmental pressures (such as climate) and from the necessity of obtaining food, the surplus of which may be directed either toward growth or toward the acquisition of new traits. In this way, the first unicellular organism emerging from the mud develops into plants; plants evolve into animals; and from animals ultimately arises the goal of the entire process: being that is able to recognise the God in the nature which turns outs to be humans (1860, 114–115, 158–165).

> In that first diminutive spark of life there was already the mark of the potential for the development of all organisms, including even the one destined to be born in the image and likeness of God. This mark, as the germ of so perfect a being as man was to become, had to develop gradually, passing through countless living workshops, enduring many struggles, and withstanding no small number of severe trials. The deep strata of the Earth are witnesses to these vital histories; and the progress in the perfection of all living beings, though it may appear inscrutable, is repeated before our eyes in spirit, as once it unfolded in the flesh. Just as each new generation of human beings lives a different life and manifests ever fresher and higher powers of spirit, so too in former times every new plant, though resembling the old, was nevertheless more perfect than it. (Chałupczyński 1860, 180)

To reconstruct and explain the process of evolution, one cannot rely on empiricism alone. For Chałupczyński, the reconstruction of humanity's natural history must take into account both empirical data and intuitive-rational explanations grounded in existing knowledge (1860, 172). On the basis of fossil remains of extinct animals (1860, 159), observations of the influence of food on the condition of organisms, and the reactions of living beings to environmental changes (1860, 252–253), one can argue that there exists a process responsible both for the transformation of one species into another and for the extinction of those unable to adapt to new conditions. The course of this process Chałupczyński deduces from his own philosophical premises. From the atomic perspective, evolution is a change in the composition of the "niedziałki," in which a new active element appears and modifies the form of the being. In this sense, there occurs among atoms a kind of "struggle for dominance," in which the determining factor is which element becomes the stimulus for the transformation of the others. Changes in atomic composition are closely connected to changes in the environment: the simpler the organism, the greater the influence of environmental factors on its structure (1860, 174–179).

Plants, as the first significant product of the evolutionary process, change primarily under the influence of environmental conditions (e.g., soil quality) and the type of nourishment available. Food serves as the source of matter and of new atoms, which can give rise to new forms. Environmental factors play a dual role: on the one hand, they determine which forms can emerge; on the other, they act selectively by eliminating those unable to survive under given conditions. The organisms that do survive continue to develop, drawing matter from those that have perished—thus reflecting Chałupczyński's hypothesis of atomic circulation—and in turn passing on traits favourable to survival to the next generation. Successive generations thereby diversify under the combined influence of inherited characteristics and changing environmental conditions (1860, 189–240, 242–243, 266–268).

> All these partial varieties of plants serve in part as a means of preserving the existence of exhausted lineages, and in part as tasks whose resolution marks the fulfilment of the plant's life-purpose insofar as it is needed as a cause for further manifestations of life. For plants do

> not live for themselves, but in order to bring forth more perfect beings; and in this respect flowers and seeds play an exceedingly important role. [...] Their ease of dispersal across wide distances, and the strong resistance with which they struggle against the most unfavourable conditions of life, are the reason for the diversification of plants on the one hand, and for the spread and durability of their existence on the other. (Chałupczyński 1860, 226)

The struggle against unfavourable environmental conditions led the first plants to a significant evolutionary leap. When some of them found themselves submerged, a new process of adaptation began, resulting in a transitional form between plants and animals—the so-called zoophyte (1860 238–239). This form was distinguished by its capacity for movement, though still passive, caused by sea currents. The rudimentary ability to move, first appearing in the zoophyte, subsequently became the essential feature distinguishing animals from plants. As Chałupczyński explains, the evolution of living organisms tended toward a gradual emancipation from environmental constraints—first through the development of independent movement, which arose in molluscs that evolved from zoophytes (1860, 242–244). It is important to emphasise that the evolution of one form into another did not imply the replacement of earlier forms but rather their diversification: older forms could continue to exist alongside newer ones, at their own level of complexity. In this way the biological diversity visible today came into being (1860, 249).

With the appearance of animals, the evolutionary process became more complex. Organisms capable of independent movement had to adapt to a greater number of factors: new environments through which they moved (1860, 320), new types of nourishment (1860, 292–299), and the presence of other organisms, with which they competed, upon which they preyed, or from which they had to defend themselves. All these factors encouraged the development of traits favourable to survival, and at the same time drove further evolution (1860, 355, 376). The mechanism of trait formation, however, remained the same: it was based on the assimilation of matter and its shaping under environmental influence. The difference was that animals possessed a volitional element—they could, to some extent, "decide" which traits to develop. Here Chałupczyński appealed to the mechanism of use and disuse. Animals sought either environments most favourable to survival—safe and nutritionally adequate—or ways of life that would allow them to neutralise adverse conditions (1860, 350–351). To achieve this, they had to cultivate those senses most useful in given circumstances. Identifying a mode of survival led to the use of the relevant organs, which, through use, underwent further development (1860, 270, 325, 328, 331, 351–352). Chałupczyński's scheme of evolutionary transformation can thus be summarised as follows: under environmental change, an organism develops its senses (in connection with modifications to the nervous system) in order to find a way to survive; it then exercises the relevant faculties, developing or modifying organs that enable it to adapt to the new conditions (1860, 328, 356). The newly developed traits are transmitted to subsequent generations, leading to the diversification of individuals. Not all traits, however, are advantageous—some may prove maladaptive and ultimately cause the death of the organism (1860, 269–270). Evolution is therefore not a simple march toward perfection, but a process in which some species perish, losing the struggle both against changing environments and against other organisms.

At this point Chałupczyński's theory comes closest to the theory of natural selection: in nature there is an unceasing struggle for survival, in which those individuals prevail whose traits have the greatest adaptive value. Like Darwin, the author of *Pomysły...* invoked Malthus to elucidate the specificity of his vision of nature.

> Thus, by necessity and by the logical sequence of law, only the greatest powers of body and of the animal soul could be transmitted widely to posterity, and not base lineages and ignoble individuals, whose mere existence could not guarantee their preservation. Here the social laws of Malthus find their sole application—with merciless benefit for life; here the numerical superiority of lower beings serves as prey for the mighty, fitting the purpose well by submitting to the minority of higher, more perfectly formed animals. Yet even among the highest there is never peace, for the struggle for dominion over a given region of the earth continues without end. This is not, however, a war of extermination or mutual annihilation, though it does not proceed without victims, but rather a relentless contest between the quantity of matter and the quality of force, between body and animal soul, between space and time. In struggling with oppositions of every kind, the development of vital powers gains much; and the temporary or enduring character of the animal's soul depends either on the actions inherited from its forebears, or on those it undertakes of its own accord. Every animal fights either to obtain nourishment, or to possess that with which it delights to remain, or in its own defence. In this way, the objects of an animal's striving are spiritual or bodily, internal or external. (Chałupczyński 1860, 376–377)

The "spiritual" qualities of which Chałupczyński writes are nothing other than new mental traits transmitted from generation to generation (1860, 381–382). In animals, the sum of these traits is referred to as instinct, which represents a kind of potentiality: it can be actualised under certain conditions, but it can also be modified if environmental determinants change (1860, 398–401). As noted above, the development of appropriate senses constitutes the first stage in the Chałupczyński evolutionary scheme. Yet for animals to make use of these senses, a sufficiently developed nervous system is required. The evolution of spiritual qualities is therefore inseparable from physical evolution (1860, 345–346). The soul—understood by Chałupczyński as the mind—guides the animal, and is itself subject to evolution. This means that traits traditionally regarded as uniquely human already have their seeds in the animal world. For example, the moral sense is a developed form of the ability to distinguish between what benefits and what endangers the organism (1860, 348). The rudiments of property rights can be seen in the ways animals store provisions for winter and build shelters (1860, 380), while the beginnings of culture may be glimpsed in their play (1860, 381).

> How, then, does man differ from the animal, when the latter too must possess knowledge of good and evil, prudence, reflection, feeling, will, even a certain creativity, and many other faculties of the soul? When, in addition, it knows sign-language, and even has a self-consciousness recognised by philosophers as the sole mark of humanity? The difference lies only in the degree of development of these powers, in the impossibility of the animal rising to that potency from which man begins to ascend—and this is the greatest and most important difference, for it can be abolished nowhere and never. (Chałupczyński 1860, 382)

According to Chałupczyński, the human being, with all its mental faculties, emerged from the animal world, and its appearance marks the culmination of the evolutionary process (1860, 353–354, 383). The only fundamental distinction between the human and the animal world, as the author emphasises, is the sense of justice, which gives humanity the chance to bring to an end the ceaseless struggle that reigns in nature (1860, 380). Justice already has its rudiments in the animal world, in the form of parental love. In human beings, however, it develops into empathy that transcends familial bonds and embraces the wider community—first in the form of patriotism, and later, at a higher stage of development, as pacifism. In this way, humankind advances spiritually, gradually approaching its ultimate goal: the embodiment of God and the preparation for the renewed coming of His kingdom (1860, 405–406). This, as mentioned earlier, entails the destruction of the world and the renewed beginning of the circulation of atoms.

## 9.4 Chałupczyński's Theory in the Polish and International Context

*Pomysły...* did not make a significant impact on the Polish publishing market, and in the first years after its release it received neither reviews nor commentary. Interest in Chałupczyński's work arose only under the influence of another treatise in natural history—Henryk Levittoux's *Zarys filozofii natury* [The Outline of the Philosophy of Nature], published in 1869. Like Chałupczyński, Levittoux placed at the centre of his theory the motion of atoms, which he saw as driving gradual transformations in reality, manifesting in the living world as a process of evolution. Levittoux's book provoked considerable controversy within Polish intellectual circles. He presented his theory as an entirely new contribution, omitting any reference to Darwin, whose works were already known among Polish intelligentsia by that time. Reviewers, with whom Levittoux engaged in long and often bitter disputes, accused him of ignorance of contemporary science and of neglecting the most important thinkers of evolutionism. In this context Chałupczyński's name began to resurface.[4] In a review published in *Przegląd Tygodniowy* [Weekly Review], Julian Ochorowicz (1869) wrote that while Levittoux's work was a positive phenomenon in Polish science—promoting the idea of species evolution and combining it with a theistic vision of the world—some of his claims merely repeated what had already been advanced by Chałupczyński. This point was developed further by Bronisław Rejchman, the leading Polish promoter of evolutionism. Stimulated by the debate surrounding Levittoux's book, he decided to review the already forgotten *Pomysły...*. Rejchman (1869) openly stated the purpose of his review: to show Polish readers that an original evolutionary work already existed in native literature—and that it was not Levittoux's *Zarys...* as its author claimed. He portrayed Chałupczyński as a genuine innovator in Polish natural philosophy, one of the first domestic naturalists to compose a work describing the history of nature from the origins of the universe to the emergence of humanity. Although Rejchman criticised some of Chałupczyński's theses—such

[4] For more on Levittoux's theory and his debates with the reviewers of his work, see Wagner 2021.

as those concerning gravity, the transmutation of elements, and the nature of light—he nevertheless recognised him in some respects as a precursor (for example in his reflections on spectral analysis). Still, he argued that Chałupczyński's reliance on deductive methods prevented him from developing his insights further. Most important of all, in Rejchman's view, were Chałupczyński's evolutionary intuitions, which largely coincided with what Darwin had proposed in the *Origin...*:

> The theory of the development of the organic world is in nuce Darwin's theory. The author assumes that animals undergo change; that the causes of these changes are food, air, water, the average temperature of the region, latitude, elevation above sea level, soil quality, and meteorological influences—and further, the use or disuse of organs (p. 269 and elsewhere). Favourable changes (says Mr. Ch., as does Darwin) are not proper to all individuals, but only to some; and hence arises the struggle for existence, and therefore also natural selection [...]. (Reichman 1869, 5)

As Rejchman observed, the theories of Chałupczyński and Darwin were so similar that, were it not for documentary evidence showing that *Pomysły...* had been submitted to the censorship office before *Origin...* was published, one might suspect plagiarism. The only essential difference, in his view, was that Chałupczyński attributed a degree of agency to animals in the evolutionary process—something Rejchman interpreted as a Lamarckian influence. Apart from this, he insisted, Chałupczyński's work was a pioneering contribution, unjustly forgotten for many years.

These comparisons did not escape Levittoux, who addressed them in an appendix to the new edition of his *Zarys...* (now retitled *Filozofia Natury* [Philosophy of Nature]) entitled *Teoria Darwina, słowo o Talesie, o Leucypie, o Arystotelesie i o pracy Pana Chałupczyńskiego* [Darwin's Theory, a Word on Thales, Leucippus, Aristotle, and the Work of Mr. Chałupczyński]. There he admitted that he had only learned of *Pomysły...* from Rejchman's review, and thus plagiarism could not be alleged[5] (Levittoux 1874, 515). He was extremely critical of Chałupczyński's theory, dismissing the idea of plagiarising such a book as absurd (1874, 517–518). He highlighted its flaws: the rejection of Newtonian theory (1874, 516); the defence of spontaneous generation, which had already been disproved by Louis Pasteur (1874, 517); and the evolutionary ideas themselves. Levittoux was himself an opponent of Darwin's theory, which he regarded as a failed extension of Lamarck's. He saw natural selection only as a mechanism of elimination, not as a means of generating evolutionary novelty (1874, 499–500). For that reason, Rejchman's assertion of an intellectual kinship between Chałupczyński and Darwin could hardly serve as a compliment in Levittoux's eyes. He argued instead that evolutionary ideas had long been present in natural philosophy—from Goethe and Erasmus Darwin (whom he incorrectly called Charles's father rather than grandfather) to Lamarck and Geoffroy Saint-Hilaire, whom he valued most highly (1874, 518). While Levittoux conceded that Chałupczyński had arrived at a theory strikingly similar to Darwin's, he wondered whether this fact testified to the truth of evolution through the struggle for existence

[5] He responded in a similar way to Ochorowicz in his reply to the latter's review (Levittoux 1869, 74).

or to the intellectual climate of the time, which favoured such ideas. In his view, the latter interpretation was the correct one:

> Mr. Chałupczyński, earlier than Darwin, wrote his work, which Darwin could not have known; and yet in Darwin we encounter thoughts that Mr. Chałupczyński, in his own sui generis manner, as we have seen, had already expressed. It is clear, then, that such are the laws of the current of intelligence: that in addressing a given question, prepared though by different paths yet embracing certain shared notions, one mind may unwittingly strike upon the same ideas as another. (Levittoux 1874, 519)

The above quotation offers a striking reversal of Darwin's claim that the "idea of evolution" was not in the air. It suggests not only that it was, but also that the intellectual currents of the age were conducive to its being formulated in precisely such a form—one that presupposed a struggle for existence. The very existence of Chałupczyński's theory challenges one of the entrenched assumptions of historians of science about the specificity of Darwin's theory of natural selection: namely, that it could have arisen only in the cultural milieu of Victorian England. Scholarly literature has often located its origins in the synthesis of British natural theology with Malthusian doctrine (e.g., Dilley 2012, Young 1985), in the tradition of English empiricism, Newtonianism, and Lyell's uniformitarianism (e.g., Gould 1996; Kauffman 2013), or in the particularities of Darwin's own social position (e.g. Bowler 2013). Some have even concluded that the theory of selection could only have been conceived within a British context (e.g., Desmond and Moore 1994; Schweber 1985). As Bowler notes (Bowler 2013, 71):

> Natural selection may have been a true theory just waiting to be discovered, but that discovery would have been made more easily in a society that encouraged the particular habits of thought associated with political individualism.

British society, shaped by the ideology of laissez-faire and the realities of the Industrial Revolution, created favourable conditions for the discovery of natural selection. By contrast, Bowler emphasises:

> The system favoured by many Continental thinkers, in which the state is an all-powerful agency directing everyone's activities, provided a closer analogy to the typological view of species. In such a model, change could only come from sudden revolution, not gradual evolution through the summation of individual activities.

However, the case of Chałupczyński shows that continental thinkers were also capable of constructing a theory closely resembling Darwin's, and almost simultaneously. Chałupczyński shared neither Darwin's cultural nor his scientific background. He was an outspoken critic of Newton (Chałupczyński 1860, 10–11, 35; 1880, 21–22); in his later writings he also attacked British individualism (Chałupczyński 1887, 108–109) and capitalism (1887, 190, 243). Their social positions were likewise starkly different: Darwin was a respected gentleman of means, whereas Chałupczyński—persecuted by the Russian authorities for his revolutionary activities—struggled to maintain stable employment. The only clear common denominator was the influence of Malthus, whom Chałupczyński later came to criticise as well (1880, 19–21).

In the above quotation, Bowler (1988) is most likely referring to his thesis concerning the existence of a specific evolutionary paradigm, which he terms "developmental." This paradigm was, according to Bowler, characteristic of nineteenth-century evolutionism, and Darwin was the only thinker able to decisively distance himself from it. Naturalists operating within the developmental paradigm were said to promote teleological conceptions of evolution, similar to those advanced by Robert Chambers in *Vestiges of the Natural History of Creation.* The question arises, however, whether Chałupczyński can be classified as a representative of developmental transmutationism Although his belief that evolutionary development tends towards the production of perfect organisms may suggest a form of teleology, *Pomysły...* does not unequivocally endorse a directed course of evolutionary transformations. According to (Chałupczyński 1860, 130–131), the direction of change is always relative to the circumstances in which atoms combine and subsequently reconfigure. Therefore, if one can speak of any universal law, it is merely the law of continuous change—itself a random process—which entails an increase in potency. Even this increase, however, is not self-evident, since Chałupczyński notes that organisms may either continue to develop or regress in their development (1860, 180–181). There are thus no exceptional species whose attainment of perfection is guaranteed (1860, 222); rather, perfection is necessary only insofar as it enables survival in the struggle with the environment and with other organisms (1860, 209). Taking into account Chałupczyński's critique of anthropocentrism (1860, 250–251), it can be argued that the ultimate goal he describes using religious terms such as "salvation" is not restricted to humanity alone. Instead, it represents a general evolutionary tendency—namely, the survival of the best-adapted—while also presupposing the existence of a universally correct mode of adaptation, one that is linked to the recognition of God's activity in nature. Analogously, in his discussion of ethical matters, Chałupczyński (1860, 397–398) maintains that free will is not subordinated to any predetermined divine plan of creation, while nevertheless affirming the existence of a proper mode of conduct that is ultimately rewarded. Thus, even when Chałupczyński speaks of the uniqueness of humanity, he does not present it as the result of any special divine privileging. Instead, he argues that the entire world is permeated by the Holy Spirit (1860, 160), while humanity is distinguished only insofar as it "[...] believes in Him, feels Him within herself, thus possesses Him, and therefore is the ultimate goal of all progress toward perfection" (1860, 160–161). On this reading, humanity functions as the ultimate goal only in a retrospective sense. Having acquired insight into the nature of the world, human beings were thereby able to survive—and it is this acquired insight, rather than any prior election, that retrospectively positions humanity as the culmination of evolutionary development.

This raises the question of which cultural and intellectual resources enabled Chałupczyński to formulate a theory that cannot be easily categorised as part of what Bowler called the developmental paradigm, and that bears such a close resemblance to Darwin's. Some of these influences had already been identified by Rejchman (1869), who pointed to the affinities of *Pomysły...* with German Naturphilosophie and with the writings of Bronisław Trentowski.

As noted earlier, Chałupczyński drew heavily on the work of Lorenz Oken, from whom he borrowed, among other things, the idea of the sphericity of the universe and certain methodological principles. He explicitly acknowledged some of these borrowings in *Pomysły...*, although there are also numerous indirect parallels. One was his manner of explaining natural processes by reconstructing the history of their creation—a cosmogonic approach also central to Oken's *Elements of Physiophilosophy* (1847, 1–2). Like Chałupczyński, Oken maintained that every component of reality was in constant motion, driven by an original "polarity" (1847, 22–23), manifest in relations among beings (for instance between celestial bodies, 1847, 53–55) and shaping their forms. The functioning of living organisms, he held, depended in part on the action of the Earth's poles (1847, 186–187). The primordial matter out of which all things arise appeared in its unstructured form as "ether" (1847, 39–40), but could assume three fundamental states: hydrogen, oxygen, and carbon (1847, 61). Here the resemblance to Chałupczyński's physics is evident, though with one crucial difference: Chałupczyński rejected the concept of ether, insisting instead on the existence of atoms (Chałupczyński 1860, 14, 333). Like Oken, he derived from primordial matter three basic elements, which in *Pomysły...* he renamed as "płyny," "lotki" and "kruszce." The most significant affinity, however, lies in their shared transmutationism. Oken's thought strongly influenced pre-Darwinian evolutionists such as Richard Owen and Geoffroy Saint-Hilaire (Breidbach 2001, 16–17), and its traces are equally visible in Chałupczyński's writings. Oken argued (1847, 185, 199–200), for example, that life arose from marine mud (Chałupczyński: from sandy mud); that processes of nutrition accounted for growth and organic change (1847, 193–194); and that the essential difference between plants and animals was the capacity for autonomous movement (1847, 199–200, 322–323), which also marked the first sign of will (1847, 319). For him too, the plant and animal kingdoms formed stages culminating in humanity (1847, 2–3), and man appeared as a manifestation of God within nature (1847, 25–26). The crucial distinction, however, is that Oken's theory was not evolutionary in the modern sense. Unlike Chałupczyński, he proposed no mechanism of evolution. Successive organic forms emerged according to the principle of recapitulation—as analogues to the functions of the human organism. The process, therefore, did not generate genuine evolutionary "novelty" (Gambarotto 2017; Breidbach 2001, 17–18).

Another important difference between the two naturalists concerned the role of God in nature. Oken likewise placed God at the centre of his philosophy, treating Him as the foundation of existence, and everything that exists as actualised divine thought (1847, 16–17). The problem lay in distinguishing the ideal (potential) from the material (actual). It was precisely this element of Oken's theory—and more broadly of German Naturphilosophie—that Chałupczyński directly attacked:

> German philosophy, in its efforts to investigate transcendental objects, often fell into the most contradictory conclusions. Oken, for instance, while speaking of eternity, infinity, and the universe, also claimed that something had emerged from nothing. It is as if one could imagine that force existed without matter, spirit without body, God without the universe; as if the Creator, existing from all eternity, had only after some time formed something out of

> nothing, as though he had grown weary for the world, for creation itself. (Chałupczyński 1860, 3–4)

For Chałupczyński, distinguishing between God's "thoughts" and "deeds" was meaningless and excessively anthropomorphic. Since man cannot know how God acts, he ought not to make assertions about His attributes at all. In this critique of German philosophy, the influence of Trentowski is particularly evident. In his polemic against Hegel, Trentowski had called for a synthesis of idealism and materialism, arguing that matter is just as eternal as spirit (Wiśniewski 2009, 43–44). Trentowski's influence is even clearer in Chałupczyński's socio-political views. For Trentowski, social progress was inseparable from the Christian principle of loving one's neighbour—without it, every form of communal life, whether religious or political, would degenerate (Starzyńska-Kościuszko 2025, 258). Chałupczyński likewise placed love and pacifism at the pinnacle of both biological and social evolution. This was a significant departure from Oken, for whom the ultimate product of civilisation was war (Breidbach 2001, 18).

Not all of Chałupczyński's intellectual debts, however, were acknowledged explicitly. In *Pomysły...* he referred several times to unnamed "scholars" who had argued for the gradual appearance of animals on Earth (e.g., 1860, 251) and pointed out that environmental changes affected the morphology of organisms (e.g., 1860, 255–256). He also presented the mechanism of use and disuse of organs, as well as the inheritance of acquired traits, as established facts. Who were these "scholars"? The most likely answer is the intellectual milieu of Moscow University, where Chałupczyński studied medicine. At that time, evolutionary ideas were being propagated by Karl F. Rul'ye. Taking the side of Geoffroy Saint-Hilaire in his dispute with Georges Cuvier, Rul'ye had, as early as the 1840s, begun to lay the foundations of an evolutionary paradigm in Russia, focusing primarily on ethology and ecology (Райков 1955; Scudo and Acanfora 1985, 736–737; Todes 1989, 107). Rul'ye promoted these ideas both in his zoology lectures—attended by students of all natural science and mathematics faculties at Moscow University (Райков 1955, 210–211)—and through popular science publications that enjoyed wide readership across Russia. It is therefore highly probable that Chałupczyński encountered Rul'ye's evolutionary views, especially since he studied in Moscow during the peak of Rul'ye's popularising activity in 1845–1847 (Райков 1955, 166–168, 206–208).[6] This influence is also suggested by striking similarities between Rul'ye's theory of evolution and that presented in *Pomysły...*. Rul'ye's theory was a blend of Lamarckism and Geoffroyism: environmental changes acted on organisms both directly and indirectly (through the modification of behaviour), altering their morphology and giving rise to new traits, which were then inherited (Райков 1955, 341–353). Rul'ye likewise saw the culmination of the evolutionary process in humankind (Райков 1955, 338), regarding human intellectual faculties as a quantitative extension of animal instinct Райков 1955, 227–230). The entire evolutionary sequence, moreover, began—as

[6] The medical course at Moscow University lasted five years, so Chałupczyński must have begun his studies in 1844 in order to receive his diploma in 1849.

in Chałupczyński's account—with plants, from which animals gradually emerged (Райков 1955, 326–329).

The influence of the Russian intellectual milieu—especially that of Rul'ye—may explain why Chałupczyński came to believe in a consensus regarding the transformation of organisms under environmental influence and the inheritance of acquired traits. While the possibility of morphological change through external factors was acknowledged even before Lamarck, it was not often connected with evolutionary ideas (Wilkins 2009). Chałupczyński, by contrast, presented his concepts as something self-evident, not requiring Darwin's "one long argument." This style likely reflected the fact that evolutionary ideas were already well established in the intellectual environment in which he operated.

## 9.5 Chałupczyński's Response to Darwinism

In 1880, likely influenced in part by the growing popularity of Darwinism in Poland as well as by the modest recognition of his own contribution to natural philosophy, Chałupczyński published *O niektórych błędach w teorii Darwina* [On Certain Errors in Darwin's Theory]. Alongside his emphasis on the originality of his own theory (1880, 15) and his praise of Darwin as an outstanding populariser of evolutionism (1880, 1–5), the work advanced a critique of natural selection—one that, paradoxically, further highlighted the similarities between the two naturalists. *Pomysły...* and *Origin...* differed not only methodologically (Darwin's empiricism vs. Chałupczyński's deductionism) but also in scope. Chałupczyński's work was comprehensive, encompassing geology, cosmology, and biology, with a particular emphasis on botany. In *Origin...*, Darwin treated botanical matters only marginally, which Chałupczyński considered a serious shortcoming. As he wrote in *O niektórych błędach...*, the struggle for existence and natural selection could not apply to plants, which did not actively shape the conditions of their existence. In their case, everything depended on the traits they already possessed; thus—he stressed—one could not speak of an "active struggle" (1880, 48, 52). Chałupczyński also criticised Darwin's analogy between artificial and natural selection. Already in *Pomysły...* he had noted that agriculture was merely an imperfect imitation of nature's processes and could not serve as the basis for scientific conclusions (1860, 189). In 1880 he reiterated the point even more forcefully, claiming that breeding was nothing more than the manipulation of existing tendencies of change, and in no way corresponded to the evolutionary process that generated those tendencies. At best, it could serve as a useful analogy for the study of heredity. If we truly wished to understand evolution, Chałupczyński argued, we should focus instead on the origins of aberrations—that is, new heritable traits (1880, 17–19).

He also objected to the mechanism of natural selection itself, arguing that in fact it explained nothing:

> The question again arises, in which part of the body was adaptation easier and more possible? Was it the adjustment of the digestive tract to the organism's needs? Or the transformation of the forelegs and neck to satisfy the stomach? This recalls the dialectical riddle: which came first, the chicken or the egg? Such are the vicious circles to which natural selection—wrongly placed and improperly understood, confused haphazardly with adaptation and the mutual dependence of organs—leads us. (Chałupczyński 1880, 26)

The problem Chałupczyński identified lay in Darwin's nearly tautological formulation of "the survival of the fittest," which, he argued, amounted to the banality: "those survive who survive." His question was: why precisely these individuals and not others? Why are conditions favourable for some, but not for others? (Chałupczyński 1880, 25). This was one of the most frequently raised issues in the early criticism of Darwinism (Wagner 2024): the question of the origin of adaptive traits. Darwin assumed the existence of natural variability within species, which provided the raw material for selection, but he was uncertain of its ultimate cause. One of his hypotheses was that environmental conditions affected the reproductive system. Chałupczyński rejected this as inconsistent with natural observations (1880, 75). For him, the key mechanisms were use and disuse of organs and the inheritance of acquired traits (1880, 14–15). Interestingly, these were processes acknowledged by Darwin himself (he discussed modification of organs through use in 1859, 131–139), particularly later in his career (cf. Darwin 1899, 388–393). This was not the only case where Chałupczyński unjustly criticised Darwin.

The author of *Pomysły...* had not fully grasped the essence of natural selection. His summary of the theory suggested that the struggle for existence meant literal combat between individuals (cf. Chałupczyński 1880, 46–47), while selection itself appeared as a conscious choice made by an organism to secure its survival (1880, 16). Hence he "corrected" Darwin by insisting that animals did not choose but were usually compelled to change their environment—in such cases, one should speak rather of "natural compulsion" than of selection (1880, 34–35). He also reproached Darwin for too narrow a conception of the "struggle for existence," stressing that alongside competition between individuals there was an even more important evolutionary struggle with environmental conditions, moderated and regulated by inherited adaptive traits (1880, 10–11, 29, 43–44, 63–64, 77–78). Struggle should not be understood merely as the elimination of weaker individuals, for it also had a developmental function (1880, 75). Paradoxically, none of these objections actually contradicted Darwin's theory—for Darwin himself (cf. 1859, 203–206, 466–467) described natural selection in similar terms. Chałupczyński's further critique revealed additional misplaced objections. He argued (1880, 24–25) that Darwin had failed to specify the level at which selection operated—species or individuals. The point was valid and was frequently raised by Darwin's contemporary critics (Wagner 2024). Yet Chałupczyński based it on the mistaken assumption that Darwin had posited sudden, complete transformations of one species into another. He explained that such a transformation could occur only in cases of hybridisation (1880, 8). In this way, he wrongly interpreted Darwin as a saltationist, whereas Darwin (1859, 189) explicitly defended gradualism. Chałupczyński also ascribed to him a kind of species realism.

In *Pomysły...* (1860, 130–133, 180–181) he himself had presented a nominalist position—seeing nature as in constant flux at every level, and treating any fixed relations among beings as accidental. Similarly, in *O niektórych błędach...* he criticised Darwin for writing of the "origin of species," when, in his view, the very category of species was nothing more than an abstraction devised for classification, not a real feature of nature (1880, 6). Once again, contrary to Chałupczyński's assumption, Darwin would in fact have largely agreed with such a nominalist understanding of taxonomic units (cf. Darwin 1859, 52).

Where, then, did Chałupczyński's misreadings of Darwin come from? The problem likely lay in the translation of *Origin...* that he relied upon. In *O niektórych błędach...* he drew primarily on the French translation by Clémentine-Augustine Royer—a version criticised both within Darwinist circles and among Francophone naturalists.[7] The controversy stemmed especially from Royer's extensive footnotes in which she frequently altered the sense of the theory itself and her expanded introduction. In it (Royer 1862, VI–XV), she cast Darwin's theory as part of a broad philosophical conflict between empiricism and mysticism. Equally problematic was her treatment of key terms. The best-known example was her translation of "selection" as "d'élection," which in English means "election" (Royer 1866, xij). This explains why Chałupczyński interpreted Darwin's selection as a kind of conscious choice. Royer's treatment of the "struggle for existence" may likewise have led him to conclude that for Darwin nature was in a state of total war. For example, where Darwin discussed the existence of electric organs in fish, Royer added a note claiming that only the strongest individuals had survived (footnote in Darwin 1862, 277–279). Similarly, in his biogeographical discussion she inserted the claim that "lower" species always become food for the more advanced (footnote in Darwin 1862, 184). Already in her introduction, Royer argued that Darwin's theory—understood as a transposition of Malthusian doctrine onto the animal world (Royer 1862, LIV–LV)—made it possible to revise the fundamental principles of society. In her view, ideas such as equality or the protection of the weak were mere prejudices of the past, to be replaced by radical individualism (Royer 1862, LVI, LXI–LXII). Using this translation, Chałupczyński could easily have concluded that Darwin was proclaiming a destructive struggle for existence and endorsing an egoistic individualism that justified exploiting the weak. In his eyes, this was not the influence of Royer herself, but rather an expression of the spirit of British culture—its economic violence toward the working class and its colonial aggression toward militarily weaker nations (Chałupczyński 1880, 21–22). Understood through the lens of Royer's translation, Darwin's theory thus appeared to him as a representation of the brutality of English civilisation (1880, 78–80).

[7] See, for example, Darwin's letter to Joseph Dalton Hooker from 1862 (Letter no. 3721) and René-Édouard Claparède's letter to Darwin from 1862 (Letter no. 3715).

## 9.6 Concluding Remarks

The question of Darwin's priority as an evolutionist was already debated during his lifetime. Even before the publication of the *Origin...*, concerns were raised about its resemblance to Alfred R. Wallace's essay *On the Tendency of Varieties to Depart Indefinitely from the Original Type*, which was read before the Linnean Society in 1858 to settle a potential dispute. On that occasion priority was awarded to Darwin (1858, 121–122), but the discussion had little immediate impact. Questions about the originality of the *Origin...* grew only after the book's success. Darwin was then accused of plagiarism by the naturalist Richard Owen and by the Scottish farmer Patrick Matthew (Darwin 1871, 11–13). Matthew's charge carried weight, for in 1831 he had outlined in *On Naval Timber and Agriculture* a conception of evolution based on the survival of better-adapted organisms. In response, beginning with the third edition of the *Origin...*, Darwin added a "Historical Sketch" listing his precursors—those who had argued for the evolutionary emergence of species and those, like Matthew and Wallace, who had anticipated natural selection. Allegations resurfaced in the 1880s, when Samuel Butler (1880), 1–17) accused him of plagiarising Jean-Baptiste Lamarck and his grandfather Erasmus Darwin, and of erasing their role in the history of evolutionism. For decades, historiography did not lend any credence to Butler's charges—seeing Darwin as the starting point in the development of evolutionary biology. Only with the professionalisation of the history of science was evolutionary thought studied systematically, producing monographs on non-Darwinian evolutionary theories, both pre-and post-1859.

Abandoning the Darwin-centred perspective allows alternative theories to appear not as "inferior versions" of Darwin's, but as original attempts to advance natural science. Chałupczyński's theory, despite limitations, represents such striking early conception. Independent of Darwin, it proposed a similar mechanism of transformation—the survival of the fittest in the struggle for existence, which was also inspired by Malthus's writings. More importantly, it challenges the conviction, reinforced by Darwin-centred historiography, that only British culture could produce a theory of selection. Figures such as Matthew and Wallace are usually cited as evolutionists who independently embraced selectionist ideas. Yet this obscures the fact that, at the same time but in a very different cultural context, a theory arose that closely resembled Darwin's, though apart from its Malthusian influence it drew on few of the same sources. Levit and Hoßfeld (2017), in their analysis of the ways in which scientific theories are formulated and received, emphasise the role played by a researcher's nationality—that is, what they term a "metaparadigm." This metaparadigm encompasses all those factors that shape how a scientist articulates their views and that are characteristic of the socio-cultural environment and intellectual tradition contributing to their national identity. If we accept that evolution by natural selection is a fact[8]—something that can be discovered through observation of nature—it should not be surprising that the manner in which this fact is articulated

[8] Thus, assuming, in accordance with Hilary Putnam's "principle of charity" (1979, 22), that the theory of selection has a referent.

depends on the intellectual tradition of the researcher who articulates it. Naturalists develop their theories within a specific historical climate and a particular intellectual tradition, drawing on metaphors and analogies that are available to them and intelligible to their intended audience. They write in accordance with prevailing intellectual fashions, referring to popular philosophical frameworks and accepted methodological standards. Chałupczyński's theory, therefore, could not have been identical to Darwin's, as he operated within a different metaparadigm. The presence in his work of metaphysical and theological themes, intertwined with physical and biological ideas, is not surprising, since this reflects the manner in which natural philosophy was practised at the time within his cultural milieu. This, however, does not diminish the fact that he also proposed an evolutionary account grounded in the survival of the best-adapted in the struggle for existence. He thus expressed an observable fact of nature using the conceptual resources available to him. Darwin did the same, albeit within a different metaparadigm—namely, that of the British empirical tradition.

This raises an open question for future research: if Chałupczyński formulated a selection theory within the context of Polish philosophy and German and Russian science, should Darwin's theory be seen as a uniquely British achievement, or as part of wider nineteenth-century intellectual currents that crossed national boundaries? If evolution by selection is an empirical fact, should we not also recognise that it may have been articulated repeatedly, albeit through the prism of different metaparadigms? On this view, selectionism may be understood as an idea that was "in the air" in the nineteenth century, taking distinct conceptual forms among thinkers operating within different cultural and intellectual contexts, as exemplified by Darwin and Chałupczyński.

## References


Bowler, Peter J. 1988. *The Non-Darwinian Revolution. Reinterpreting a Historical Myth*. Baltimore: The Johns Hopkins University Press.

Bowler, Peter J. 2013. *Darwin Deleted. Imagining a World without Darwin*. Chicago: The University of Chicago Press.

Breidbach, Olaf. 2001. Oken in der Wissenschaftsgeschichte des 19. Jahrhunderts. In *Lorenz Oken (1779–1851). Ein politischer Naturphilosoph*, ed. Olaf Breidbach, Hans-Joachim Fliedner, and Klaus Ries, 15–32. Weimar: Hermann Bohlaus Nachfolger.

Browne, Janet. 2010. Making Darwin: Biography and the Changing Representations of Charles Darwin. *The Journal of Interdisciplinary History* 40 (3): 347–373.

Butler, Samuel. 1880. *Unconscious Memory: A Comparison between the Theory of Dr. Ewald Hering and the "Philosophy of the Unconscious" of Dr. Edward von Hartmann*. London: Ballantyne Press.

Chałupczyński, Adam. 1860. *Pomysły do wiedzy dziejów życia świata*. Warszawa: Nakładem Autora.

Chałupczyński, Adam. 1880. *O niektórych błędach w teorii Darwina*. Warszawa: W drukarni S. Orgelbranda Synów.

Chałupczyński, Adam. 1887. *Gawędy ekonomiczno-społeczne*. Włocławek: Nakładem i drukiem Hermana Neumana.

Darwin, Charles. 1859. *On the Origin of Species by Means of Natural Selection, or the Preservation of Favoured Races in the Struggle for Life*. London: John Murray.

Darwin, Charles. 1862. *L'Origine des espèces.* Translated by Clémence Augustine Royer. Paris: Librairies de Guillaumin et Cie et de Masson et Fils.
Darwin, Charles. 1871. *On the Origin of Species by Means of Natural Selection, or the Preservation of Favoured Races in the Struggle for Life.* 5th ed. New York: D. Appleton & Company.
Darwin, Charles. 1899. *The Variation of Animals and Plants under Domestication.* Vol. 2. London: John Murray.
Darwin, Charles. 1958. *The Autobiography of Charles Darwin 1809–1882.* London: Collins.
Darwin Correspondence Project. Letter No. 3715. Available at/from https://www.darwinproject.ac.uk/letter/?docId=letters/DCP-LETT-3715.xml.Accessed on 2 August 2025.
Darwin Correspondence Project. Letter No. 3721. Available at/from https://www.darwinproject.ac.uk/letter/?docId=letters/DCP-LETT-3721.xml. Accessed on 2 August 2025.
Desmond, Adrian, and James Moore. 1994. *Darwin: The Life of a Tormented Evolutionist.* New York: W.W. Norton & Company.
Dilley, Stephen. 2012. Charles Darwin's use of theology in the origin of species. *The British Journal for the History of Science* 45 (1): 29–56.
Dorcz, Krzysztof. 2016. Doktor Adam Chałupczyński, zapomniany pozytywista. *Tu i Teraz. Gazeta Włocławka i Powiatu Włocławskiego* 80: 14.
Gambarotto, Andrea. 2017. Lorenz Oken (1779–1851): Naturphilosophie and the Reform of Natural History. *The British Journal for the History of Science* 50 (2): 329–340.
Gould, Stephen Jay. 1996. *Time's Arrow, Time's Cycle: Myth and Metaphor in the Discovery of Geological Time.* Cambridge, MA: Harvard University Press.
Kauffman, Stuart A. 2013. Evolution Beyond Newton, Darwin, and Entailing Law: The Origin of Complexity in the Evolving Biosphere. In *Complexity and the Arrow of Time*, ed. Charles H. Lineweaver, Paul C. W. Davies, and Michael Ruse, 162–190. New York: Cambridge University Press.
Kucharzewski. Jan. 1914. *Epoka paskiewiczowska. Losy oświaty.* Warszawa: Skład Główny w Księgarni Gebethnera i Wolffa.
Levit, Georgy S., and Uwe Hoßfeld. 2017. Major Research Traditions in Twentieth-Century Evolutionary Biology: The Relations of Germany's Darwinism with Them. In *The Darwinian Tradition in Context. Research Programs in Evolutionary Biology*, ed. Richard G. Delisle, 169–193. Cham: Springer.
Levittoux, Henryk. 1869. Polemika naukowa. *Przegląd Tygodniowy Życia Społecznego, Literatury i Sztuk Pięknych* 9:72–75.
Levittoux, Henryk. 1874. *Filozofia Natury.* Warszawa: W drukarni Gazety Lekarskiej.
Mayr, Ernst. 1982. The Growth of Biological Thought: Diversity, Evolution, and Inheritance. Cambridge, MA: The Belknap Press of Harvard University Press.
Mościcki, Henryk. 1949. *Szymon Konarski.* Warszawa: Państwowy Instytut Wydawniczy.
Neuser, Wolfgang. 2001. Methodische Grundzüge von Lorenz Okens Lehrbuch der Naturphilosophie. In *Lorenz Oken (1779–1851). Ein politischer Naturphilosoph*, ed. Olaf Breidbach, Hans-Joachim Fliedner, and Klaus Ries, 210–216. Weimar: Hermann Bohlaus Nachfolger.
Ochorowicz, Julian. 1869. Przegląd literatury polskiej. *Przegląd Tygodniowy Życia Społecznego, Literatury i Sztuk Pięknych* 5:36–37.
Oken, Lorenz. 1847. *Elements of Physiophilosophy.* London: Printed for The Ray Society.
Putnam, Hilary. 1979. *Meaning and the Moral Sciences.* London: Routledge & Kegan Paul Ltd.
Rejchman, Bronisław. 1869. Pomysły do wiedzy dziejów życia świata napisał Adam Chałupczyński, Warszawa 1860 r. *Gazeta Polska (dawniej Codzienna)* 100:5.
Royer, Clémence Augustine. 1862. Preface du traducteur. In *Charles Darwin, L'Origine des espèces.* Paris: Librairies de Guillaumin et Cie et de Masson et Fils.
Royer, Clémence Augustine. 1866. Avant-Propos. In *Charles Darwin, L'Origine des espèces*, i–xiij. Paris: Librairies de Guillaumin et Cie et de Masson et Fils.
Schweber, Silvan S. 1985. The Wider British Context in Darwin's Theorizing. In *The Darwinian Heritage*, ed. David Kohn, 35–69. Princeton: Princeton University Press.

Scudo, Francesco M., and Michele Acanfora. 1985. Darwin and Russian Evolutionary Biology. In *The Darwinian Heritage*, ed. David Kohn, 731–752. Princeton: Princeton University Press.
Seyda, Bronisław. 1999. *Słownik encyklopedyczny lekarzy pisarzy w literaturze światowej*. Szczecin: Wydawnictwo Promocyjne "Albatros".
Starzyńska-Kościuszko, Ewa. 2025. Bronisław Ferdynand Trentowski—jako filozof polityki i polityczny cybernauta. *Humanistyka i Przyrodoznawstwo* 29:253–285.
Todes, Daniel P. 1989. *Darwin without Malthus. The Struggle for Existence in Russian Evolutionary Thought*. New York: Oxford University Press.
Wagner, Michał. 2021. Evolutionary Theism of Henryk Levittoux in Polish and International Context. *Kwartalnik Naukowy Fides et Ratio* 48 (4): 53–73.
Wagner, Michał Jakub. 2024. *Revisiting the Eclipse of Darwinism. A Historiographical and Philosophical Analysis*. Cham: Springer.
Wilkins, John S. 2009. *Species. A History of the Idea*. Berkeley: University of California Press.
Wiśniewski, Radosław. 2009. W poszukiwaniu cech filozofii polskiej—A. Cieszkowski, B. F. Trentowski i K. Libelt. *Hybris* 8 (1): 30–58.
Young, Robert M. 1985. *Darwin's Metaphor: Nature's Place in Victorian Culture*. Cambridge: Cambridge University Press.
Zawadzki, Arkadiusz. CHAŁUPCZYŃSKI Adam (1822–1887), powstaniec styczniowy, lekarz. In [online] *Słownik biograficzny Południowego Podlasia i Wschodniego Mazowsza*. https://slownik-biograficzny.uws.edu.pl/slownik/588-chalupczynski-adam-1822-1887-powstaniec-styczniowy-lekarz. Accessed 25 August 2025
Райков, Борис Евгеньевич. 1955. *Биологи-Эволюционисты До Дарвина Материалы К Истории Эволюционной Идеи В России*. Vol. 3. Москва: Издательство Академии Наук Ссср.

**Michał Jakub Wagner** is a faculty member at Cardinal Stefan Wyszyński University in Warsaw. His academic background is in the philosophy of nature, as well as the philosophy and history of science. His principal research interests include the reception of Darwinism in the nineteenth century, non-Darwinian theories of evolution, and problems in the development of science. He is the author of Revisiting the Eclipse of Darwinism: A Historiographical and Philosophical Analysis.

# Chapter 10
# Wittgenstein's Critique of Science and Technology

**Moritz René Pretzsch**

**Abstract** In this paper, I examine Wittgenstein's critical thinking on science and technology. I will argue that Wittgenstein's fundamental anti-scientism and anti-technicism can still be fruitfully applied today, that his scepticism towards the causal nexus is timelessly important, and that he approached the inexplicable with a certain awe and respect. I will also argue that a meaningful and instructive parallel can be drawn between Wittgenstein's late thinking and Thomas Kuhn's concept of paradigm shift. Approaches to this can be found in Wittgenstein's concept of world-picture. What is more, Wittgenstein's problem of so-called 'deep disagreement' may also offer an interesting position for the natural sciences. If we look carefully, we can see that Wittgenstein already offered approaches to tackling the problem of incommensurabilities and bringing about an exchange between different (scientific) world-pictures. Finally, I will discuss how Wittgenstein's pursuit of self-knowledge and his attitude of humility can be understood as important for our crisis-ridden present. Wittgenstein's thinking is still timeless and extremely beneficial for the natural sciences.

**Keywords** Wittgenstein · Science · Technology · Scientism · Reason · Rationality · Causal nexus · Kuhn · Modernity · Culture · Progress

## 10.1 Introduction

In this article, I would like to clarify the following question: What is Wittgenstein's critique of modernity? In search of an answer to the first question, I will focus my attention, after a brief introduction, on Wittgenstein's critique of science and technology, and in doing so I will also discuss his view of what is unavoidable. In a concluding reflection, I will explain what our scientistic and technicistic age can learn from Wittgenstein's critique of modernity. Wittgenstein had planned to publish

M. R. Pretzsch (✉)
Department of Philosophy, University of Kassel, Kassel, Germany
e-mail: moritzpretzsch@uni-kassel.de

 
E. Jung et al. (eds.), *The Forgotten Giants: Overlooked Revolutions in Natural Philosophy*, Studies in History and Philosophy of Science 70,
https://doi.org/10.1007/978-3-032-34008-5_10

notes that were included in the complete edition of Wittgenstein's writings under the title *Philosophical Remarks* with a preface, of which there are several versions. A version of the preface from November 1930 states:

> This book is written for such men as are in sympathy with its spirit. This spirit is different from the one which informs the vast stream of European and American civilisation in which all of us stand. That spirit expresses itself in an onwards movement, in building ever larger and more complicated structures; the other in striving after clarity and perspicuity in no matter what structure. The first tries to grasp the world by way of its periphery—in its variety; the second at its centre—in its essence. And so the first adds one construction to another, moving on and up, as it were, from one stage to the next, while the other remains where it is and what it tries to grasp is always the same. I would like to say 'This book is written to the glory of God,' but nowadays that would be chicanery, that is, it would not be rightly understood. It means the book is written in good will, and in so far as it is not so written, but out of vanity, etc., the author would wish to see it condemned. (Wittgenstein 1975, Preface)

The tendency of European and American civilisation is alien and unsympathetic to Wittgenstein (cf. Pretzsch 2025a, 279–288). A civilisation that is scientific and technological in nature and essentially characterised by progress: a series of structures, progress for progress' sake, the growth of complicated, confusing structures, a striving away from the centre into the diversity of the 'periphery,' a loss of clarity. In particular, Wittgenstein considers the belief in progress—one of the cornerstones of Western civilisation since the Enlightenment—which manifests itself in a trust in instrumental reason, to be particularly responsible for the cultural decline of his era. The spirit of this striving for progress is diametrically opposed to Wittgenstein's own thinking. For him, it is superficial, clinging to the periphery, failing to penetrate deeply and understanding nothing of the essence of the world. The optimistic machine age is in danger of losing itself in superficial diversity. Scientific knowledge is growing rapidly. Proportional to this growth, the overview is diminishing. In other words, for Wittgenstein, the programme of modern rationality—particularly as pursued within Western science—is typically characterised by its constructive orientation (cf. Wittgenstein 1980a, 7 and see Pretzsch 2025b). The "ever more complicated structure" (Wittgenstein 1980a, 7), what is constructed in this process is, for him, a construction, an achievement made possible by technology which, precisely because it is heteronomously motivated, cannot find satisfaction in itself, cannot constitute an "end in itself" (Wittgenstein 1980a, 7). For Wittgenstein, the 'spirit' of this civilisation, the idea of progress through scientific and technological mastery, is therefore a mindset of appropriation, a strategy of expansion, one-upmanship and the unrestrained increase of instrumental rationality. It is therefore no wonder that Wittgenstein does not favour the "typical western scientist" (Wittgenstein 1980a, 7), representing this spirit of escalating logic. Wittgenstein explicitly mentions that he philosophises in a completely different spirit than scientists pursue their sciences and that the latter cannot understand him (cf. Wittgenstein 1980a, 7). Hardly any other modern philosopher has made such a strong attempt to reconcile existence and his own way of thinking as Wittgenstein.

What is more, Wittgenstein always strove to combine the solution of his philosophical problems with the redemption of his own existence. For him, philosophical problems are life problems. Similar to Ray Monk, one can take the view that Wittgenstein's philosophical quest and his emotional and intellectual life form a unity (cf. Monk 1990, xviii). In this respect, we cannot overemphasise the importance of positioning his thinking, as found, for example, in the draft preface of 1930. Wittgenstein's thinking essentially seeks clarity, transparency and overview, and does so with a radicalism and uncompromisingness that is unparalleled. More importantly, clarity, transparency and overview are ends in themselves for him (cf. Wittgenstein 1980a, 7).[1] For this reason, too, Wittgenstein cannot write in his draft preface that his book 'was written to the glory of God,' since such a non-purposeful investigation would seem incomprehensible to the dominant spirit of his era.[2] Wittgenstein remarked to a friend: "My type of thinking is not wanted in this present age, I have to swim so strongly against the tide. Perhaps in a hundred years people will really want what I am writing." (Drury 1984, 79) His genius harbours, in Habermas' words, a "feeling of isolation, of living in an uncomprehending, quasi-categorically biased environment." (Habermas 2017, 21 f.; my translation) According to Habermas, Wittgenstein, like Heidegger and Adorno, addresses scattered individuals who are willing to be initiated into a new way of thinking, rather than the communication community of researchers or the cooperation community of citizens (cf. Habermas 2017, 21 f.). And just like Heidegger and Adorno, Wittgenstein also turns away from institutionalised science and political public life (cf. Habermas 2017, 21 f.). "The philosopher is not a citizen of any community of ideas. That is what makes him into a philosopher." (Wittgenstein 1967, 81)—states Wittgenstein in one of his writings. Wittgenstein's thinking is essentially out of step with the times.

## 10.2 Wittgenstein's Critique of Science and Technology

Wittgenstein's critique of modernity is, to a large extent, a critique of science. Wittgenstein was by no means hostile to science, but he opposed the scientistic spirit in which scientists in European and American civilisation typically practise and instrumentalise science. Two main points of criticism can be identified in Wittgenstein's critique of science. First, he opposes the fantasies of omnipotence, hubris, and the obsession with explanation that pervade scientific and technological civilisation, as exemplified by scientism[3]—in short, the idea that everything can be explained scientifically. Second, Wittgenstein takes issue with the naivety and carelessness of many representatives of the scientific community, highlighting their lack of insight

[1] With his programme of "clarity as an end in itself," argues Matthias Kroß, Wittgenstein commits himself "to the philosophical habitus of contemplative observation as a way of viewing the world that is not oriented towards ends but towards itself." (Kroß 1993, 32; my transl.).

[2] For Matthias Kroß, God can be conceived for Wittgenstein as 'the causa sui, as the epitome of clarity as an end in itself' par excellence (cf. Kroß 1993, 32 and see in this context Krebs 2025, 323–336; Munz 2025, 289–304 and Somavilla 2025, 305–322).

[3] Wilhelm Vossenkuhl puts it succinctly: "The entire structure of his thinking compels him to oppose scientism." (Vossenkuhl 1995, 169; my transl.). See also Vossenkuhl 2012, 295–307.

and self-awareness. Wittgenstein simply wants to advocate for 'humility in science' (cf. McGuinness 2017, 9),[4] as McGuinness puts it succinctly.

### *10.2.1 First Main Point of Criticism*

Regarding the first main point of criticism: a central element of Wittgenstein's critique of the scientific-technical civilisation permeated by scientism is his critical examination of the causal nexus in the *Tractatus Logico-Philosophicus.* "Superstition is nothing but belief in the causal nexus" (Wittgenstein 2002, 5.1361),[5] states Wittgenstein in the *Tractatus Logico-Philosophicus.* Belief in the causal nexus is the belief that our understanding of the world is based on scientific explanations of causal relationships. Superstition is this belief when we think we can use it to say something about the world as a whole. Superstition is particularly evident when we also think we can know something real about the world in a scientific manner without having to be aware of the metaphysical character of this belief. Wittgenstein's critique is thus not aimed at the natural sciences as such, but rather at the implications this insight has for the scientific worldview—specifically, the belief that science provides an exhaustive account of reality. What he opposes is the ideological transformation of scientific thought that has come to dominate modern scientific and technological civilisation, namely scientism. In the *Tractatus Logico-Philosophicus*, he writes: "The whole modern conception of the world is founded on the illusion that the so-called laws of nature are the explanations of natural phenomena." (Wittgenstein 2002, 6.372)[6] It is the illusion that the world is explained as soon as we know the laws of nature according to which the phenomena in the world come into being, change and pass away. Consequently, one sentence later in the *Tractatus Logico-Philosophicus*, he says the following about the representatives of the modern worldview:

> Thus people today stop at the laws of nature, treating them as something inviolable, just as God and Fate were treated in past ages.
> And in fact both are right and both wrong: though the view of the ancients is clearer in so far as they have a clear and acknowledged terminus, while the modern system tries to make it look as if *everything* were explained. (Wittgenstein 2002, 6.372)

Scientific and technological civilisation believes that everything can be explained by the laws of nature. For it, the laws of nature are the ultimate reality. But it is mistaken.

When one considers the world from a logical perspective, as Wittgenstein does in the *Tractatus Logico-Philosophicus*, natural laws are by no means fundamental; rather, it is the truth-functions of the elementary propositions that occupy this role.

---

[4] See also Habermas's insightful interpretation of Wittgenstein's anti-scientism: Habermas 2017, 22 f.

[5] As standard practice, I cite Wittgenstein's *Tractatus Logico-Philosophicus* (*Tractatus*) (TLP) by sentence number rather than page number.

[6] At this point, Wittgenstein could certainly have written 'modern European and American civilisation' or even 'scientific and technological civilisation.'

The truth-values of these elementary propositions determine the truth of all other propositions with logical necessity, including that of the natural laws. Beyond this logically necessary, truth-functional determination of all true propositions by the elementary propositions, there exists no independent causal or natural-law necessity. But then, according to Wittgenstein, one should not speak of something being natural or causally possible, impossible or necessary. Wittgenstein writes in the *Tractatus Logico-Philosophicus*:

> There is no compulsion making one thing happen because another has happened. The only necessity that exists is *logical* necessity. (Wittgenstein 2002, 6.37)

Of course, the truth values can be distributed among the elementary propositions in a different way than they actually are. Such a different distribution would represent a logically possible alternative to our world. Again, the laws of nature that would apply in this alternative world would be logically necessary. However, according to the *Tractatus Logico-Philosophicus*, whether or not a different distribution of truth values among the elementary propositions is possible is decided solely from the standpoint of logic, not from the standpoint of the laws of nature such as the law of causality. A noteworthy consequence emerges: the world as a whole is determined in purely logical terms by which elementary propositions are true and which are false. Yet the question persists as to why the elementary propositions possess precisely these truth-values rather than others. No definitive answer can be given to this question. Logic is incapable of supplying such an answer, since its domain encompasses all possible configurations of truth and, for that very reason, cannot privilege or exclude any particular one. From the logical standpoint, it is therefore a matter of contingency which elementary propositions are true. This contingency marks a fundamental limit of logical explanation and delineates the boundary between what can be said meaningfully and what can only be acknowledged as given. At this boundary, explanation gives way not to further justification, but to the recognition of the sheer factuality of the world. The exploration of logic means the exploration of *everything that is subject to law*. And outside logic everything is accidental. (Wittgenstein 2002, 6.3).

It is crucial to emphasise in this context that neither the law of causality nor individual laws of nature are capable of explaining the truth or falsity of elementary propositions. This is because the laws of nature themselves are truth-functions of elementary propositions and therefore presuppose, rather than determine, their truth-values. Consequently, the world as a whole cannot be explained—neither by logic nor by natural laws. There is no explanation for why this particular world exists rather than another, nor for why there is a world at all. From a logical perspective, it is not nothingness but the existence of the logical space of possible worlds that calls for explanation (see also Sect. 10.3). Yet such an explanation is impossible, since it would have to be formulated from a standpoint outside the logical space of all possible worlds and thus beyond logic itself—a standpoint that cannot be attained.

The ultimate point of reference, when the world is considered from a logical standpoint, is therefore the logical space of all possible distributions of truth-values among elementary propositions. While a given distribution determines which laws of nature hold in a particular case, the laws of nature themselves cannot, conversely, account for why certain elementary propositions are true rather than false. Only from this logical standpoint does it become clear what can and cannot be explained, and that the laws of nature do not genuinely explain anything—not individual facts, nor even the world understood as the totality of facts within logical space.

Those who, like "the ancients" (Wittgenstein 2002, 6.372) in pre-modern eras, say that God created the world and only God knows why he created it, put an end to all explanations, but also reveal that we humans cannot explain why there is a world at all and why this world exists rather than any other. Scientific and technological civilisation—for which causality is *the* paradigm of modern, scientific explanation of the world—can explain none of this with the laws of nature, yet gives the opposite impression. For this reason alone, it should not mock "the ancients" (Wittgenstein 2002, 6.372). And there is another important point: in its fantasies of omnipotence, its scientific obsession with explanation and its hubris, the scientific-technological civilisation of our time fails to realise that the most important thing, the thing that concerns us most, our life problems, is not even touched upon by any scientific explanations of the world (cf. Wittgenstein 2002, 6.52). What, according to Wittgenstein, are our life problems? To do good, to live in harmony with the world, to "live happily."[7] Science has no answer to what it means to live this way and what we must do to live this way. The modern idea that, in principle, everything can be explained on the basis of natural laws is, for Wittgenstein, nothing more than hubris, an attempt to 'take the place of God.' For him, it is a pitiful attempt to paint over and forget the enigmatic unavailability of the world and life. Scientistic pseudo-answers and supposed further explanations must remain unsatisfactory. The belief that the world can be explained once the facts have been scientifically explained is, for Wittgenstein, an illusion. The findings of natural science are misused for ideological and worldview purposes. What is more, the natural sciences themselves are idolised, which he sees as the clearest symptom and also an important cause of cultural decline (cf. Monk 1990, 299).[8] This is the first main point of Wittgenstein's critique of scientific and technological civilisation. And he does not deviate from this later on, even after he has abandoned the doctrine of elementary propositions. In his later work in particular, he returns to the question of causality when he examines the "groundlessness of our believing" (Wittgenstein 1969, 166) in our worldview. Just like the religious worldview, the scientific worldview also lacks an ironclad, absolutely secure foundation. According to Vossenkuhl, Wittgenstein's entire philosophical work attempts to make clear the absence of secure foundations in all areas of life, thought and belief, and to convince us that we are not missing anything because of this (cf. Vossenkuhl 1995, 301).[9] However, a scientistic worldview is reluctant to admit the absence of a secure foundation. In his search for a pre-established order, Wittgenstein relentlessly

[7] See in particular his diary entries on 6 July 1916; 8 July 1916; 14 July 1916; 29 July 1916; 30 July 1916; 2 August 1916; 13 August 1916; 20 October 1916 and 21 October 1916. See Wittgenstein 1990.

[8] This is also reminiscent of Oswald Spengler, who believed that when a culture decays, the once living organism solidifies into a dead, mechanical structure. A creative, artistic era is followed by the reign of mathematics and mechanics.

[9] See also the study by Nida-Rümelin 2012, 271–294 in this context.

exposes the groundlessness of our worldview, even if our scientific and technological civilisation is reluctant to see this. Ultimately, it is also the scientific method, in its pursuit of generality, that Wittgenstein sees as the source of problems, which he counters with his concept of family resemblance, among other things.[10]

### *10.2.2 Second Main Point of Criticism*

On to the second main point of criticism: Wittgenstein's criticism of science and technology is particularly clear, alongside his criticism of scientific and technological delusions of omnipotence, hubris and the obsession with explanation, in relation to the irresponsible events of the time: the dropping of atomic bombs on Hiroshima and Nagasaki. Wittgenstein sees something good in the atomic bomb, at least insofar as he sees it as a possible end to science perverted by specialisation. He speaks about this on 19 August 1946:

> The hysterical fear over the atom bomb now being experienced, or at any rate expressed, by the public almost suggests that at last something really salutary has been invented. The fright at least gives the impression of really effective bitter medicine. I can't help thinking: if this didn't have something good about it the *philistines* wouldn't be making an outcry. But perhaps this too is a childish idea. Because really all I can mean is that the bomb offers a prospect of the end, the destruction, of an evil,—our disgusting soapy water science. And certainly that's not an unpleasant thought; but who can say what would come *after* this destruction? The people now making speeches against producing the bomb are undoubtedly the *scum* of the intellectuals, but even that does not prove beyond question that what they abominate is to be welcomed. (Wittgenstein 1980a, 48f.)

But just one year later, on 14 July 1947, Wittgenstein had already given up this hope. Instead, he saw how the form of science he despised was becoming established in its connection with industry, a development that not only continues to this day but is accelerating:

> Science and industry, and their progress, might turn out to be the most enduring thing in the modern world. Perhaps any speculation about a coming collapse of science and industry is, for the present and for a *long* time to come, nothing but a dream; perhaps science and industry, having caused infinite misery in the process, will unite the world—I mean condense it into a *single* unit, though one in which peace is the last thing that will find a home.
>
> Because science and industry do decide wars, or so it seems. (Wittgenstein 1980a, 63)

Wittgenstein's scepticism towards science, technology, scientism and the belief in progress of his time culminates in his view of time in 1947. It states:

> It isn't absurd, e.g., to believe that the age of science and technology is the beginning of the end for humanity; that the idea of great progress is a delusion, along with the idea that

[10] Elsewhere, with regard to the scientific pursuit of generalisation, he states: "Science: enrichment and impoverishment. One particular method elbows all the others aside. They all seem paltry by comparison, preliminary stages at best. You must go right down to the original sources so as to see them all side by side, both the neglected and the preferred." (Wittgenstein 1980a, 61)

> the truth will ultimately be known; that there is nothing good or desirable about scientific knowledge and that mankind, in seeking it, is falling into a trap. It is by no means obvious that this is not how things are. (Wittgenstein 1980a, 56)

This, too, is reminiscent of Spengler and his idea that modern 'Faustian man' has 'become a slave to his own creation' (cf. Spengler 1990, 1190), enslaved by the machine, which is 'only now unfolding its diabolical secret power' (cf. Spengler 1990, 1191). He remarks to his friend and student Norman Malcom:

> Only, extraordinary scientific achievements have a way these days, of being used for the destruction of human beings (I mean their bodies, or their souls, or their intelligence). So hold on to your brains. (Malcolm 1984, 39)

Looking at our crisis-ridden times, many of Wittgenstein's prescient thoughts seem to be coming true. What's more, we can benefit from his warnings if we take them seriously and apply them to our own time. In this context, I find Wittgenstein's idea that science and technology must redefine themselves particularly helpful: freed from orthodox elementary thinking, in which only the threat of punishment or the prospect of reward guarantees compliance with rules and laws without ever questioning them. Results obtained in this way are then either right or wrong, free of any ethical or moral baggage, and if the possibilities obtained in this way lead to conflicts with and within our world, society may intervene with prohibitions and commandments. On the other hand, science and technology that are responsible and reasonable in their actions do not need 'ethics committees' to control them. The necessary moral and ethical criteria are derived from science and technology themselves, and only in this way are they effective. The perspective must be changed, a different approach, a different relationship to the world must be learned. What is needed is a new way of looking at the world through the lens of science and technology.

Wittgenstein wrote about what this new responsible approach might look like on 2 September 1916, when he reflected on the meaning and position of humans in the world:

> The human body [...] *my* body in particular, is part of the world among other parts of the world among animals, plants, stones, etc., etc.
>
> Anyone who understands this will not want to give their body or the human body a privileged place in the world.
>
> They will naively regard humans and animals as similar and belonging together. (Wittgenstein 1990, 177; my transl.)[11]

Overall, Wittgenstein's criticism of science and technology aims to undermine a way of thinking that he considers partly responsible for the horrors of the two world wars and the Holocaust—in short, for industrialised killing in the twentieth century. A way

[11] Diary entry from 2 September 1916. My translation. The original German text reads: „Der menschliche Körper [...] *mein* Körper ins Besondere, ist ein Teil der Welt unter anderen Teil der Welt unter Tieren, Pflanzen, Steinen etc. etc.

Wer einsieht, wird seinem Körper oder dem menschlichen Körper nicht eine Bevorzugte Stelle in der Welt einräumen wollen.

Er wird Menschen und Tiere ganz naiv als ähnliche und zusammengehörige Dinge betrachten."

of thinking that blindly pursues progress, in which it may be difficult for individuals to strive for a life worth living and to cooperate peacefully with others. Ultimately, it is a way of thinking that one can only hope will come to a complete end.

## 10.3 Wittgenstein and His View of the Inexplicable

In his *lecture on ethics*, Wittgenstein states that modern man finds satisfaction solely in the fulfilment of one of the most 'despicable desires' of all: the fulfilment of his "superficial curiosity about the latest discoveries of science" (Wittgenstein 1965, 4).[12] Wittgenstein also criticises science for becoming a sedative that lulls people to sleep, robbing them of their capacity for wonder, since this requires them to be wide awake.[13] The ability to marvel, which since Plato has been the very approach to philosophy, means not being satisfied with the ordinary and the self-evident, and encountering existence and the world with reverence and decency. It also means overcoming nonsensical ambitions to explain things where they are out of place. In modern times, people try to eliminate or simply ignore the unavoidable, the inexplicable, the unavailable. But what is the unavoidable, the unavailable and inexplicable that makes people marvel? For Wittgenstein, since his early work, the *Tractatus Logico-Philosophicus*, this is what exists beyond the world of facts, beyond everything that is the case: the transcendent space.[14] This is where we must locate the mystical, the absolute, meaning, aesthetics and ethics. In other words, the mystical is the miracle *that* something *is* and not nothing. That is, it is the miracle *that* the world is (cf. Wittgenstein 2002, 6.44). According to Wittgenstein's philosophy of language, it is certainly true that we cannot speak meaningfully about this: "Only something supernatural can express the Supernatural." (Wittgenstein 1980a, 3) And yet this does not mean that it does not exist. Nor should we lose our relationship to it. Wittgenstein sums this up well when he says:

> It is very *remarkable* that we should be inclined to think of civilization—houses, trees, cars, etc.—as separating man from his origins, from what is lofty and eternal, etc. Our civilised environment, along with its trees and plants, strikes us then as thought it were cheaply wrapped in cellophane and isolated from everything great, from God, as it were. That is remarkable picture that intrudes on us. (Wittgenstein 1980a, 50)

---

[12] And see his remark in the lectures: "Jeans has written a book called *The Mysterious Universe* and I loathe it and call it misleading. Take the title. This alone I would call misleading. [...] I might say the title *The Mysterious Universe* includes a kind of idol worship, the idol being Science and the Scientist." (Wittgenstein 1967, 27) Elsewhere he says: "I may find scientific questions interesting, but they never really grip me. Only *conceptual* and *aesthetic* questions do that. At bottom I am different to the solution of scientific problems; but not the other sort." (Wittgenstein 1980a, 79)

[13] See, for example, the remark: "Man has to awaken to wonder—ad so perhaps do peoples. Science is a way of sending him to sleep again." (Wittgenstein 1980a, 5)

[14] See also in this context: Pretzsch 2023c and Pretzsch 2023e, 476–848.

The sphere of influence of science and technology is solely the world of facts. Within this sphere, it considers objects solely from the perspective of their control and under the premises of their usability and usefulness. Problems arise precisely when this mastery—as discussed above—becomes a worldview, and the idea arises that nothing can escape scientific and technological endeavour, their power of disposal. In other words, when one follows the delusion of considering only that to be real which is available to rational cognition working with the methods of mathematical and physical science. (cf. Bultmann 1952, 289) It is the narrow-minded belief in science as a panacea that Wittgenstein is sceptical about. In 1941, he writes:

> What a curious attitude scientists have: "We still don't know that; but it is knowable and it is only a matter of time before we got to know it!" As if that went without saying. (Wittgenstein 1980a, 40)

The insight into the limits of knowledge, as painful as it may be, is completely lost here, as Wittgenstein outlines. It is important to accept the inexplicable as inexplicable with willingness and wonder, and not simply to deny it out of wounded cognitive vanity. In short, it is important to take people's metaphysical needs seriously.[15] We must recognise that not everything can be explained scientifically and controlled technically. Humans, as Wittgenstein repeatedly points out, must admit that despite their supposed self-empowerment, despite their scientific, technical and cultural achievements, they remain bound by their dependence, powerlessness and limitations, not least through the experience of their finitude. Human beings cannot explain *that* they exist, *that* the world exists. As Wittgenstein showed in his *Lecture on Ethics* and under the influence of Kierkegaard (cf. Pretzsch 2023a, 149–160; Pretzsch 2023b, 2023c), certain human experiences cannot be clearly formulated in a scientific and meaningful way. They are simply not accessible to the language of facts, and yet they are of existential importance: wonder, security, guilt. What makes life worth living is of absolute value. However, such things cannot be found in the world and, according to Wittgenstein, cannot be meaningfully represented through language. The crucial point is that Wittgenstein, like Kant, protects the transcendent space through his imperative of silence in the *Tractatus Logico-Philosophicus*[16] and thus also from false claims of knowledge and esoteric chatter. To those who nevertheless wish to speak about it, we can echo St Augustine's call: 'Be silent and understand.'

---

[15] Even if we cannot articulate this metaphysical need in a meaningful way, this does not mean that it does not exist for Wittgenstein.

[16] Thus, the famous last sentence in the *Tractatus*—"What we cannot speak about we must pass over in silence" (Wittgenstein 2002, 7)—contains both a logical-philosophical truth and an ethical imperative (cf. Pretzsch 2023d, 105–128 and Pretzsch 2024, 628–636). Wittgenstein's critique of the misuse of language continues in his *Philosophical Investigations* and has recently been recognised on several occasions for its cultural-critical implications.

## 10.4 Wittgenstein and a Look at Thomas Kuhn

Wittgenstein himself and his thinking were an important sounding board for new scientific progress, for example when it came to the changes surrounding intuitionism in mathematics. His late philosophy of language in particular offers seismographic possibilities for recognising when a new era of scientific progress begins and ends. This is particularly evident in the scientific terms that human beings hold on to and consider important. With each new era of scientific progress, a tremor runs through the world picture and the language of human beings. When one paradigm is replaced by another, the language associated with it also changes (cf. Wittgenstein 1969, 96–97; Wittgenstein 1980b, 103–104). Old technical terms no longer apply and no longer accurately reflect new scientific findings. Either their old meaning must be changed or new technical terms must be coined. But even before a new era dawns, doubts about potentially outdated paradigms can be reflected in language. How can this be understood in more detail? New theories do not arise in principle as soon as the limits of the prevailing theory are reached, but emerge as drafts by individuals that are rejected by the majority of experts in the field. It is precisely these individuals, the pioneers, who develop their own language, so to speak, and play new language games. The paradigm shift begins not least with language. All this immediately brings to mind the philosophy of Thomas Kuhn.[17] Although there are already numerous research papers that deal intensively with Wittgenstein and Kuhn's thesis of incommensurability (cf. Gunnell 2014), there are hardly any that examine how language and concepts, according to Wittgenstein, introduce, represent and manifest new paradigms. This is still uncharted territory. Wittgenstein himself sometimes used the term 'paradigm,'[18] but of course in a different sense than Kuhn. But let us first take a closer look at Wittgenstein's concept of world picture. I will then turn to Thomas Kuhn's paradigm shift.

Wittgenstein refers to the concept of world-picture as "a system of what is believed" (Wittgenstein 1969, 144). A major "difficulty" now lies in realising "the groundlessness of our believing." (Wittgenstein 1969, 166) For what is believed cannot be proven to be adequate to reality (cf. Majetschak 2019, 123). Thus, what we believe is not based on reality, but rather determines what is to be considered reality in the first place (cf. Majetschak 2019, 123). This also means that there is no way to prove the truth of the basic assumptions of any world-picture. The question, "But is there then no objective truth?" (Wittgenstein 1969, 108) is ultimately meaningless and philosophically naive, because

> [a]ll testing, all confirmation and disconfirmation of a hypothesis takes place already within a system. And this system is not a more or less arbitrary and doubtful point of departure for all our arguments: no, it belongs to the essence of what we call an argument. The system

---

[17] And Kuhn himself mentioned Wittgenstein several times in his work (see, for example: Kuhn 1996, 44 f.).

[18] The earliest references can be found in Wittgenstein's estate from the 1930s onwards. Including minor variations in manuscripts, the term 'paradigm' appears twenty times in Wittgenstein's estate.

> is not so much the point of departure, as the element in which arguments have their life. (Wittgenstein 1969, 105)

For Wittgenstein, there can be no truth or reason applicable to humans that is abstracted from all possible reference systems, just as there can be no meta-perspective or world-picture-transcending standard. This is because justifications and rationalisations for beliefs and truths are only valid *within* a world-picture, "but justification comes to an end" (Wittgenstein 1969, 192) at the latest when we encounter the so-called 'ultimate reasons' at the centre of a world-picture. Belief in a world-picture is holistic (cf. Wittgenstein 1969, 141 f., 225) and does not allow for any standpoint external to language games or world-pictures that could be considered neutral. For this reason, it is also impossible to identify any reasons, standards or norms that are authoritative within a world-picture as true, correct or binding in a sense that transcends all world-pictures. Knowledge, truth or reason are valid—as Wittgenstein's late philosophical approaches show—only within a world-picture. And everything that can be considered true or unquestionable knowledge is "in the end based on acknowledgement." (Wittgenstein 1969, 378) This acknowledgement can only ever take place within the specific parameters of one's own world-picture. And this applies not only to the concept of knowledge, but also to truth and reason.

Wittgenstein's concept of world-picture further implies that it is problematic to label the standards of other world-pictures as "primitive" from one's own perspective. Such a judgement ultimately means absolutising one's own world-picture. However, we lack the universally accepted standard by which truth or falsehood can be objectively measured, and such a standard cannot exist. Furthermore, we must bear in mind that we always use "our language game" and "fight" (cf. Wittgenstein 1969, 609) the other when we describe the basic beliefs of another world-picture as "primitive" or "false" from the perspective of our own. Once again, the following applies here: we can only talk about other world-pictures in our own language and on the basis of our own world-picture. However, it is uncertain whether so-called "deep disagreement" (Fogelin 1985 and see Godden and Brenner 2010; Pritchard 2018), i.e., the debate between different world-pictures, leads to anything at all. As Wittgenstein famously remarked at one point: "Where two principles really do meet which cannot be reconciled with one another, then each man declares the other a fool and heretic." (Wittgenstein 1969, 611) Even if there can be no meta-rationality when two world-pictures encounter each other, there is a widespread strategy—frequently used in the fight against other world-pictures—which, as Wittgenstein recognised without illusion, consists of supporting one's own approach with all kinds of slogans (Wittgenstein 1969, 610) and, for example, claiming the concept of 'reason' for one's own arguments. Thus, one might say to the representative of the foreign world-picture, 'Every reasonable person must admit that...', thereby excluding those who insist on different beliefs from the community of reasonable people. However, this widespread argumentation strategy overlooks the fact that what we call "reasonable" can only be determined in the light of a world-picture. This shows that we cannot argue with members of different world-pictures, but this does not mean that we are released from our world-picture-internal standards of rationality in such a situation.

Although we are forced to resort to strategies of 'persuasion' and 'conversion' (cf. Wittgenstein 1969, 262, 612)—insofar as peaceful interaction is possible at all—nothing allows us to be irrational in doing so. Even the methods and content used in the rhetorical strategies mentioned above remain open to discussion and justification within the horizon of our own world-picture. Deception or the threat of violence in the event of a "deep disagreement" are in no way tolerable. Instead, we can start with fundamental beliefs that the other person shares and try to change their world-picture from there through stories, demonstrations, experiments, etc.

Let us take a look at Thomas Kuhn's thinking at this point. It is no exaggeration to say that no book on science in the twentieth and twenty-first centuries has had a greater impact on the discussion than Kuhn's *The Structure of Scientific Revolutions*, first published in 1962. Kuhn not only changed the way experts think about science. With his approach, Kuhn successfully rejected the long-dominant view of science known as the *received view*. According to Kuhn, science proceeds in four phases: the pre-paradigmatic phase, the normal science phase, the phase of anomalies and crises, and the revolutionary phase, which brings with it potential incommensurabilities. Let us focus briefly on the first two phases:

The first is the so-called pre-paradigmatic phase. Strictly speaking, science has not yet begun to be science in this phase. This phase is characterised by the fact that no doctrine has yet established itself in the respective field, meaning that there are usually as many different schools of thought as there are scientists involved in the debate. Kuhn cites optical theories in the pre-Newton era as an example of a pre-paradigmatic phase. The substantive work that is characteristic of pre-paradigmatic phases has little to do with the usual scientific debates. Pre-paradigmatic scientists are not concerned, for example, with determining the value of a natural constant to an additional decimal place or searching for an entity postulated by theory. Instead, pre-paradigmatic scientists are engaged in fundamental discussions that primarily address the question of what research in the discipline in question should look like. The debate therefore focuses on fundamental ontological questions, questions regarding the appropriate methods of data collection, or questions regarding basic epistemic standards. In order to decide questions such as these, pre-paradigmatic scientists do not shy away from using unconventional methods of reasoning such as philosophical arguments or thought experiments. The typical pre-paradigmatic publication begins with a plea for its own way of describing scientific research—but it rarely goes beyond these programmatic first steps. Of course, it is not entirely impossible that genuine scientific achievements can also be made in pre-paradigmatic phases. However, the probability of concrete solutions in individual scientific fields is low because almost all energy is devoted to trying to convince other scientists of the merits of one's own vision of scientific work. According to Kuhn, science in the true sense of the word only begins once the pre-paradigmatic phase has been overcome and normal scientific research begins. Scientists suddenly do everything they did not do (or at least not systematically) in the pre-paradigmatic phase: they abandon fundamental discussions and instead try to determine the value of a natural constant to an additional decimal place or to prove the existence of an entity postulated by a theory. In this context,

Kuhn famously refers to scientists as 'puzzle solvers' (cf. Kuhn 1996, 35–42). Pre-paradigmatic scientists are, in a sense, like players who are primarily concerned with convincing each other of the merits of their preferred rules of the game. Normal scientific players, on the other hand, differ from this in that they have agreed on certain rules of the game and recognise these as binding for their practical actions. This comparison naturally raises an obvious question: What is it that pre-paradigmatic scientists do not share, but normal scientists do? In other words, what is it that, according to Kuhn, is as constitutive for science as rules are for games? Kuhn's answer is that normal scientists are characterised by a common paradigm, in much the same way that players in a game are characterised by common rules.

The similarity between Wittgenstein's and Kuhn's approaches becomes clear when we consider that paradigms are associated with a certain way of thinking. According to Kuhn, we can understand a paradigm as the totality of recognised scientific achievements, assumptions, values, methods and concepts shared by a scientific community in a particular discipline. When a paradigm reaches its limits due to contradictions or new findings, a scientific revolution and paradigm shift may occur, in which a new paradigm replaces the old one and changes science. In Wittgenstein's words, one could say that behind a paradigm there is an entire frame of reference, a world-picture, and that a paradigm expresses a style of thinking. A style of thinking that is conveyed through numerous language games. When a paradigm shift is imminent, there is no absolute standard by which the 'correctness' or 'incorrectness' of the respective paradigm can be measured. According to Wittgenstein's approach, such a standard does not exist and is not even possible. It would be completely nonsensical to say that one world-picture is correct and another is incorrect, because how could this be measured? A paradigm is also 'correct' as long as it is not replaced by another and thus corrected. Members of the Newtonian paradigm considered their paradigm to be just as 'correct' and 'conclusive' as the members of the Einsteinian paradigm later did. But there is no absolute basis for measurement. Instead, the field is left to *the* style of thinking and its language games, and finds its appeal in offering more orientation, more support, more plausibility, and appearing more meaningful to human beings at a particular time.

## 10.5 Wittgenstein and the Present Day

What can we conclude from the previous sections? In my opinion, the core of Wittgenstein's critique of modernity is his reservation towards any intense form of world domination.[19] This is precisely what makes it so relevant and topical in our crisis-ridden times. World-empowerment refers to any form of appropriation and control over the world. It refers to the attempt to bring the world or significant parts of it under a certain rule, control or influence. At first glance, there is nothing wrong with this, as world-domination forms the basis of every culture and civilisation that a human being laboriously wrests from nature. And there is no doubt that important achievements in world-domination—just think of medicine—have brought about a fundamental reduction in suffering and an increase in prosperity. It becomes explosive, as Wittgenstein astutely and presciently observed in his time, and as we see in our own time, when world-domination has taken on excessive and (self-)destructive traits. When a human being's sense and understanding of every form of dependence, power and limitation has been completely lost. When progress exists only for the sake of progress and, in a sense, runs on empty. World-domination has then crossed a line, and its out-of-control endeavours are destroying human beings and the environment.

In the course of the unstoppable progress of their fundamentally scientific and technological civilisation, modern human beings have adopted the involuntary expectation that everything will always happen as they have calculated and planned. This applies both to individual life plans and to political and economic forecasts that affect the community. For a long time, the credo was that 'we have it in our hands' if we can only muster the will and the means for our purposes: For the future no longer 'comes'; we no longer understand it as the 'coming'; we make it (cf. Anders 1985, 313). The self-image of the modern human being, intoxicated by the unstoppable pursuit of progress of the optimistic scientific and technological civilisation, is completely misguided. Life in the modern age is inscribed with a logic of growth that exhibits world-domination and is fundamentally never satisfied with 'enough.' And yet the modern human being has reached his own limit, where he is smaller than himself, can no longer keep up with himself, that is, can no longer cope with his products (cf. Anders 1985, 47). He can no longer keep up with 'himself' and his technical achievements. In a sense, they are always one step behind and can never fully clarify the later consequences of their actions, yet they must live with the consequences of their actions. And so it is in our time: human beings have become

[19] The metaphorical breadth of the term world domination is notable: it encompasses not only technological control or scientific rationality, but also the broader ethical, existential, and societal consequences of humanity's drive to dominate nature and restructure the world. Without this clarification, it is unclear whether one is referring solely to technological mastery, scientific rationality, or the broader cultural ethos that underpins these developments. In Wittgenstein's view, unchecked world domination undermines humility, self-awareness, and the recognition of human limitations, producing a civilization in which progress is pursued for its own sake and humans risk becoming both the authors and the victims of their own creations. His philosophical project, with its relentless pursuit of clarity, thus functions as a corrective, emphasising reflection, ethical awareness, and a measured understanding of the human condition.

the authors of their own catastrophe. In other words, in the consequences of their own world-domination, human beings' supposed power of disposal is destroyed. As the active creator of his achievements, he now finds himself passively exposed to them. In his unbridled accumulation of wealth, in which all sense of proportion has been lost, he sometimes even forgets his own inescapable finitude. And in doing so, he also forgets that there may be something beyond his all-encompassing desire for power and his thirst for domination. Blind world-domination also means denying the unavailable and the inescapable.

It is precisely these varieties of arrogant and self-aggrandising thinking that Wittgenstein's critique of modernity is directed against. His entire philosophy is essentially aimed at clarity. Clarity about one's own relationship to oneself and the world. Wittgenstein does not merely quibble a little with the explanatory concepts of scientific and technological civilisation. No, the radical and fundamental nature of his criticism lies in the fact that he sees fantasies of omnipotence at work, which must be rejected as fundamentally ethically wrong. The modern belief in progress could not accommodate humility as the recognition of one's own dependence, powerlessness and limitations. Modernity lacks a selfless recognition of reality (cf. Murdoch 2001, 95) and of its own limitations. To put it briefly and simply: modernity lacks humility. Or, in the words of Holm Tetens: Philosophy in the spirit of Wittgenstein is not a doctrine. It seeks to practise humility (cf. Tetens 2020, 156 and see also Krebs 2025, 323–336; Munz 2025, 289–304 and Somavilla 2025, 305–322). In our time, we can understand Wittgenstein and his thinking as exemplary and guiding for our actions. We, too, must constantly, again and again, take up the work on ourselves in order to find our way back to a correct and healthy self-image. In short, Wittgenstein's attitude of humility is the right corrective to world-domination as we need it today. It forms the basis for a reflection that openly problematises the character of human endeavour. Those who do not want to approach their environment solely against the backdrop of world-domination with a view to profitable gain can, with humility and openness, remain aware of their dependence, powerlessness and limitations and put a stop to their arrogant and destructive endeavours. The continuing interest in Wittgenstein's thinking to this day suggests that we have recognised his problems as the problems of our time.

## References

Anders, Günther. 1985. Die Antiquiertheit des Menschen. In *Über die Seele im Zeitalter der zweiten industriellen Revolution*, vol. 1. München: C. H. Beck.

Bultmann, Rudolf. 1952. *Die Bedeutung des Gedankens der Freiheit für die abendländische Kultur*. Tübingen: Mohr Siebeck.

Drury, M. O. C. 1984. Some Notes on Conversations with Wittgenstein. In *Recollections of Wittgenstein*, ed. Rush Rhees, 76–96. Oxford: Oxford University Press.

Fogelin, Robert J. 1985. The Logic of Deep Disagreements. *Informal Logic* 7 (1): 1–8.

Godden, David M., and William H. Brenner. 2010. Wittgenstein and the Logic of Deep Disagreement. Cogency. *Journal of Reasoning and Argumentation* 2 (2): 41–80.

Gunnell, John G. 2014. *Social Inquiry After Wittgenstein and Kuhn: Leaving Everything as it is*. New York: Columbia University Press.

Habermas, Jürgen. 2017. Gedanken bei der Vorbereitung einer Konferenz. In *Der Löwe spricht... und wir können ihn nicht verstehen—Ein Symposium an der Universität Frankfurt anläßlich des hundertsten Geburtstags von Ludwig Wittgenstein*, ed. Brian McGuinness et al., 20–26. Frankfurt am Main: Suhrkamp.

Krebs, Andreas. 2025. "[I]ch sehe jedes Problem von einem religiösen Standpunkt" Gibt es eine religiöse Fortschrittskritik bei Wittgenstein? In *Special Topic: Das erlösende Wort in dürftigen Zeiten—Wittgensteins Fortschritts-, Zivilisations-und Kulturkritik*, ed. Moritz René Pretzsch, 323–336.

Kroß, Matthias. 1993. *Klarheit als Selbstzweck—Wittgenstein über Philosophie, Religion, Ethik und Gewißheit*. Berlin: Akademie.

Kuhn, Thomas. 1996. *The Structure of Scientific Revolutions*. Chicago: University of Chicago Press.

Majetschak, Stefan. 2019. *Wittgenstein und die Folgen*. Stuttgart: Metzler.

Malcolm, Norman. 1984. *Ludwig Wittgenstein. A Memoir*. Oxford: Oxford University Press.

McGuinness, Brian. 2017. Der Löwe spricht... und wir können ihn nicht verstehen.'Zu Wittgensteins hundertstem Geburtstag. In *Der Löwe spricht... und wir können ihn nicht verstehen—Ein Symposium an der Universität Frankfurt anläßlich des hundertsten Geburtstags von Ludwig Wittgenstein*, eds. Brian McGuinness et al., 7–19. Frankfurt am Main: Suhrkamp.

Monk, Ray. 1990. *Ludwig Wittgenstein: The Duty of Genius*. London: Random.

Munz, Volker A. 2025. Text und Kontext: Wittgenstein und seine Zeit. In *Special Topic: Das erlösende Wort in dürftigen Zeiten—Wittgensteins Fortschritts-, Zivilisations-und Kulturkritik*, ed. Moritz René Pretzsch, 289–304.

Murdoch, Iris. 2001. *The Sovereignty of Good*. London: Taylor & Francis Ltd.

Nida-Rümelin, Julian. 2012. Lebensform, Philosophie und Wissenschaft—eine Wittgensteinsche Perspektive. In *Wittgenstein: Zu Philosophie und Wissenschaft*, ed. Pirmin Stekeler-Weithofer, 271–294. Hamburg: Felix Meiner.

Pretzsch, Moritz René. 2023a. Surprising Parallels in the Understanding of the World in the Early Works of Ludwig Wittgenstein and Martin Heidegger. In What Can Be Shown Cannot Be Said. *Proceedings of the International Ludwig Wittgenstein Symposium, Zagreb, Croatia*, 2021, ed. Ines Skelac and Ante Belić, 145–156. Münster: LIT.

Pretzsch, Moritz René. 2023b. *Selbstwerdung und Selbstverfehlung*. Baden-Baden: Verlag Karl Alber.

Pretzsch, Moritz René. 2023c. *Das Weltverständnis im Frühwerk von Martin Heidegger und Ludwig Wittgenstein*. Baden-Baden: Tectum Verlag.

Pretzsch, Moritz René. 2023d. Erlösung durch Ästhetik. Schopenhauers Befreiung der Erkenntnis von der Herrschaft des Willens durch die ästhetische Kontemplation und eine Bezugnahme zum ästhetischen Denken des frühen Wittgenstein. *Schopenhauer-Jahrbuch* 104:105–128.

Pretzsch, Moritz René. 2023e. "Erfahrung" und Erfahrung—eine Untersuchung des Satzes 5.552 im *Tractatus Logico-Philosophicus*. In *100 Years Tractatus. Beiträge des 44. Internationalen Wittgenstein Symposiums*, ed. Esther Heinrich-Ramharter et al., 476–484. Kirchberg am Wechsel: Österreichische Ludwig Wittgenstein Gesellschaft.

Pretzsch, Moritz René. 2024. Der *Tractatus Logico-Philosophicus* und die Rede vom Verstehen. In *Facetten der Wirklichkeit—Zeitgenössische Debatten. Beiträge des 45. Internationalen Wittgenstein Symposiums*, edited by Yannic Kappes et al., 628–636. Kirchberg am Wechsel: Österreichische Ludwig Wittgenstein Gesellschaft.

Pretzsch, Moritz René. 2025a. Einleitung. In *Special Topic: Das erlösende Wort in dürftigen Zeiten—Wittgensteins Fortschritts-, Zivilisations-und Kulturkritik*, ed. Moritz René Pretzsch. *Wittgenstein-Studien* 16(1): 279–288.

Pretzsch, Moritz René, ed. 2025b. Special Topic: Das erlösende Wort in dürftigen Zeiten—Wittgensteins Fortschritts-, Zivilisations-und Kulturkritik. *Wittgenstein-Studien* 16(1).

Pritchard, Duncan. 2018. Wittgensteinian Hinge Epistemology and Deep Disagreement. *Topoi* 40:1117–1125.

Somavilla, Ilse. 2025. Wozu Philosophie und Kunst in Zeiten der Unkultur? In *Special Topic: Das erlösende Wort in dürftigen Zeiten—Wittgensteins Fortschritts-, Zivilisations-und Kulturkritik*, ed. Moritz René Pretzsch, 305–322.
Spengler, Oswald. 1990. *Der Untergang des Abendlandes*. München: C. H. Beck.
Tetens, Holm. 2020. *Wittgensteins "Tractatus"—Ein Kommentar*. Stuttgart: Reclam.
Vossenkuhl, Wilhelm. 1995. *Ludwig Wittgenstein*. München: C. H. Beck.
Vossenkuhl, Wilhelm. 2012. Wittgensteins Wissenschaftskritik. In *Wittgenstein: Zu Philosophie und Wissenschaft*, ed. Pirmin Stekeler-Weithofer, 295–307. Hamburg: Felix Meiner.
Wittgenstein, Ludwig. 1965. A Lecture on Ethics. *The Philosophical Review* 74 (1): 3–12.
Wittgenstein, Ludwig. 1967. *Zettel*, eds. G.E.A. Anscombe and G.H. von Wright, trans. G.E.M. Anscombe. Oxford: Basil Blackwell.
Wittgenstein, Ludwig. 1969. *On Certainty*, eds. G. E. A. Anscombe and G. H. von Wright, trans. Denis Paul and G. E. M. Anscombe. Oxford: Basil Blackwell.
Wittgenstein, Ludwig. 1975. *Philosophical Remarks*, ed. R. Rhees. New York: Barnes & Noble, trans. Raymond Hargreaves and Roger White.
Wittgenstein, Ludwig. 1980a. *Culture and Value*, ed. G. H. von Wright in collaboration with Heikki Nyman, trans. Peter Winch. Chicago: The University of Chicago Press.
Wittgenstein, Ludwig. 1980b. Wittgenstein's Lectures: Cambridge, 1930–1932. In *From the Notes of John King and Desmond Lee*, ed. D. Lee. Oxford: Basil Blackwell.
Wittgenstein, Ludwig. 1990. *Tagebücher 1914–1916*. Frankfurt am Main: Suhrkamp.
Wittgenstein, Ludwig. 2002. *Tractatus Logico-Philosophicus*. London: Routledge.

**Moritz René Pretzsch** is a Research Associate in the Department of Philosophy at the University of Kassel. His research focuses, among other areas, on philosophy of language, epistemology, metaphysics, philosophy of culture, philosophy of religion, aesthetics and phenomenology. He has held numerous research appointments and visiting positions abroad, including extended stays in Norway, Finland, Denmark, and Japan. Pretzsch is a member of the Executive Board of the International Ludwig Wittgenstein Society. His recent publications include monographs on Søren Kierkegaard, Ludwig Wittgenstein, and Martin Heidegger, addressing questions of rationality, language, and self-knowledge. He also serves as co-editor, together with Jan Kerkmann, of the book series Fundamental Concepts of Human Existence and Rationality and Criticism of Rationality, and together with Anna-Lisa Scherfose of the series Terms of Art.

# Chapter 11
# Maria Ossowska and the Science of Morality: A Forgotten Giant of Experimental Moral Philosophy?

**Joseph Ulatowski**

**Abstract** Maria Ossowska wrote extensively on moral philosophy, and much of her work has focused on the foundations of moral philosophy, including its psychological and sociological aspects. From the turn of the twenty-first century a branch of experimental philosophy has investigated moral intuitions, judgements, and behaviours using social scientific methods and data to substantiate, challenge, or ameliorate claims of moral philosophy. When we look at Ossowska's contributions to what she called the science of morality, we cannot help but see the overlapping similarities between her own research programme and the empirical work that has been undertaken by moral psychologists and experimental philosophers. In my contribution to this volume, I aim to compare Ossowska's vision of the science of morality with experimental moral philosophy and to argue that given the commonalities between them, Ossowska is a Forgotten Giant whose contribution to philosophy paved the way for moral psychology and experimental moral philosophy to become what it is today.

**Keywords** Maria Ossowska · Experimental philosophy · History of philosophy · Lvov-Warsaw school

## 11.1 Introduction

Maria Ossowska has a vast body of work that spans across the twentieth century and in a variety of philosophic subdisciplines from philosophy of language to moral philosophy. She was among the first scholars from Poland to write about the philosophy of Bertrand Russell (Kosecki, forthcoming; Bednarz 2022; Woleński 2022). Ossowska published works as a solo author, as well as collaboratively with her spouse, Stanisław Ossowski. Despite the importance of her philosophical work, she

J. Ulatowski (✉)
University of Waikato, Hamilton, Aotearoa, New Zealand
e-mail: joe.ulatowski@waikato.ac.nz



E. Jung et al. (eds.), *The Forgotten Giants: Overlooked Revolutions in Natural Philosophy*, Studies in History and Philosophy of Science 70,
https://doi.org/10.1007/978-3-032-34008-5_11

remains an enigmatic figure in the history of analytic philosophy, a forgotten giant of the field.

Her major treatises in moral philosophy include *Podstawy nauki o moralności* (*The Foundations of the Science of Morality*, 1947), *Motywy postępowania—Z zagadnień psychologii moralności* (*Human Motivation—An Inquiry into the Psychology of Moral Life*, 1949), and *Moralność mieszczańska* (*Bourgeois Morality*, 1956), among others some of which have appeared in English in the 1970s and 1980s. Much of her work focused on the foundations of moral philosophy, including its psychological and sociological aspects.

From the turn of the twenty-first century a branch of moral philosophy has investigated moral intuitions, judgements, and behaviours using social scientific methods and data to substantiate, challenge, or ameliorate claims of moral philosophy. It has been called "experimental moral philosophy". When we look at Ossowska's contributions to what she called "the science of morality," we cannot help but see overlapping similarities between her own research programme and work in experimental moral philosophy. While Ossowska may not have conducted experiments herself, the spirit of her work anticipates the value of what experimental moral philosophers would eventually undertake at the end of the twentieth and beginning of the twenty-first century. It would seem from the extant literature available in English that Ossowska would have very much appreciated the development of empirical methods to serve philosophical enquiry.

In this contribution, I aim to compare Ossowska's vision of the science of morality with experimental moral philosophy and to argue that given the commonalities between them, Ossowska is a *forgotten giant of experimental moral philosophy* whose contribution to philosophy paved the way for moral psychology and experimental moral philosophy to become what it is today. In §11.2, I provide a gloss of the *long* history of experimental philosophy. I do so because few people working in and around the contemporary version of experimental philosophy appreciate that it is not a methodology that sprung up suddenly at the end of the last century. On the contrary, the historical roots of experimental philosophy are quite deep. Once I have shown that experimental philosophy has a long and deep history, I turn to the moral philosophy of Ossowska and argue in §11.3 that her work exemplifies what we have come to expect from experimental philosophy. Her empirical work, I contend, may be due in part to her intellectual upbringing in Twardowski's Lvov-Warsaw School where methodological innovation in philosophical projects was lauded rather than derided. In §11.4, I address four potential challenges that a critic may raise against my view of Ossowska, especially that I am imposing an idiosyncratic view of my own making on hers. In opposition, I contend that it is the critic whose view that has been unjustly imposed upon Ossowska. Once I have addressed these concerns, I conclude in §11.5 that Ossowska was a pioneer and forgotten giant of experimental moral philosophy.

## 11.2 Experimental Philosophy and Its Long History

A group of philosophers emerged in the early 2000s that appeared to be doing something new. These philosophers did not rely upon their own claims about what non-philosophers believed, and they did not limit themselves to reporting the results published in scientific papers that they believed were relevant to their philosophical aspirations. Instead, they took on the role of conducting systematic empirical research themselves. Their goal was to enrich philosophical discussions through the collection of empirical data that could help address philosophical questions. In this section, I define experimental philosophy and highlight an early example of this seemingly novel phenomenon in 21st-century philosophy. The example serves as excellent entry points for those new to contemporary experimental philosophy.

Experimental philosophy ("x-phi" for short) is one of the newest tools in the philosopher's toolkit, and its main function is to test the veracity of empirical facts (*cf.* Barnard et al, 2021). X-phi is a way of *doing* philosophy that uses empirical methods, such as questionnaires or surveys, corpus analyses, or neuroscientific methods (e.g., fMRIs, EEGs, etc.), to inform and enrich one's position in a philosophical debate. There is not just one way of doing x-phi; instead, we should think of it as an abundant suite of experimental techniques. However, as Barnard et al (2021) have argued, there is not just one means of engaging in x-phi; on the contrary, there are different ways in which x-phi may be undertaken. And, even if the researcher does not herself perform experiments or empirical studies, this does not exempt them from having done philosophy experimentally.

For purposes of clarity and continuity, let me tighten up that definition by rehearsing a definition first introduced by Joshua Knobe and Shaun Nichols (2008, 2017):

Experimental philosophy: a means of doing research that uses empirical methods, usually employed by linguistics, psychology, and the social sciences, to investigate the use of philosophically salient concepts in natural language or to explore the psychological underpinnings of philosophical problems.

Research in x-phi thus "brings together two key elements":

1. the kinds of questions and theoretical frameworks traditionally associated with philosophy; and
2. the kinds of experimental methods traditionally associated with psychology and cognitive science. (Knobe and Nichols 2017, 1)

So, for research in philosophy to be a piece of x-phi, its focus should be a philosophical problem and the methods the research uses should be those that are commonly employed by the social sciences.

While I mentioned that practitioners of x-phi use a diverse range of methods and experimental design, one popular experimental design derived from social and cognitive psychology uses questionnaires to survey respondents about their intuitions. For instance, Joshua Knobe's CEO harm and help scenarios test research participants about their non-reflective views of intentional action (Knobe 2003; 2004; cf. Adams and Steadman 2004a; 2004b). He gave each study participant one of the two scenarios and asked them to answer the question whether the chairman of the board intended to harm/help the environment. Knobe discovered an interesting asymmetry where respondents seemed to have been moved to decide whether an action is intentional, not based upon the action itself but the moral valence of the consequences of the action, i.e., the side-effect of the action. Here are the two scenarios:

Harm scenario:

> The vice-president of a company went to the chairman of the board and said, 'We are thinking of starting a new program. It will help us increase profits, but it will also harm the environment.'
>
> The chairman of the board answered, 'I don't care at all about harming the environment. I just want to make as much profit as I can. Let's start the new program.'
>
> They started the new program. Sure enough, the environment was harmed.

Help scenario:

> The vice-president of a company went to the chairman of the board and said, 'We are thinking of starting a new program. It will help us increase profits, and it will also help the environment.'
>
> The chairman of the board answered, 'I don't care at all about helping the environment. I just want to make as much profit as I can. Let's start the new program.'
>
> They started the new program. Sure enough, the environment was helped.

88% of study participants responded that the chairman in the harm scenario intentionally harmed the environment, while 77% of study participants said that the chairman in the help scenario did *not* intentionally help the environment. That is a strikingly odd result not only because it is contrary to what many action theorists have thought about intentional action but also because the only difference between the two scenarios is in one instance the environment was harmed and in the other it was helped.

Questionnaires may seek to uncover individual differences, whether it be by gender, sexual identity, socio-economic status, education, philosophical sophistication, or ethnicity. Other experiments may use more sophisticated technology like scans from fMRI (functional magnetic resonance imaging), EEG (electroencephalography), and NIRS (near-infrared spectroscopy) to analyse neural activity while study participants perform tasks, such as being presented with a trolley case (cf. Greene et al. 2001). Recently, experimental philosophers have begun using corpus linguistics to draw conclusions about philosophical questions (e.g., Hansen et al 2021; Sytsma et al 2019). The aim of these studies is to collect and analyse existing linguistic data on the use of words (Biber et al 1998; McCarthy and O'Keefe 2010; McEnery and Wilson 2001). This aim is pursued using *corpora*, which are collections of written or oral texts that are typically curated to give a balanced and representative picture of the target domain of language use.

The label "experimental philosophy" was commonly employed in the early modern period but then had largely fallen into disuse by the end of the nineteenth century, and one may suspect that the use of empirical methods in philosophy largely died with the label as philosophy increasingly came to be distinguished from the sciences. However, a more likely story is the following. Philosophy generated disciplinary offspring that more deftly addressed specialised research questions using a variety of empirical methods. As this process of specialisation continued, theoretical speculation increasingly came to overshadow empirical investigation within philosophy. Despite this, it is important to recognise that empirical work never disappeared from Western philosophy. As we will see in this section and the next, regardless of the standard story about early analytic philosophy, a clear lineage can be found between nineteenth- and twentieth-century philosophy and contemporary x-phi.

A standard story of the rise of analytic philosophy centres around the increased focus on logico-linguistic issues by the likes of Gottlob Frege and Bertrand Russell. Despite the impact that some philosophers had in the emerging scientific psychology, and despite inroads being made using empirical methods by philosophers, their overwhelming influence turned many philosophers' gaze toward language, ontology, and the foundations of mathematics. Accordingly, the view that predominated analytic philosophy from the 1880s through the early part of the twentieth century was that formal and speculative approaches, in contrast to empirical or experimental approaches, form the basis of resolving philosophical problems (cf. Dummett 1993).

However, this story of early analytic philosophy's relationship with empirical work is incomplete. There are many instances of people that we typically categorise as American philosophers who conducted experiments and made careful empirical observations to advance their philosophical (and psychological) theses. For example, while Charles Sanders Peirce and William James often published in philosophy journals, some of their work was published in early issues of *Psychological Review* and *The American Journal of Psychology*. Similarly, Henri Bergson's first scholarly publication (1886) was based on his observations of hypnosis sessions, and he was elected president of the *Society for Psychical Research* in 1913. Moreover, John Dewey, Josiah Royce, and William James each served as President of the American Psychological Association around the turn of the century. The end of the nineteenth and beginning of the twentieth century was rife with philosophical activity in psychology labs, as we illustrate over the next several paragraphs.

Even a cursory glance at philosophy journals such as *The Philosophical Review* or *Mind* and psychology journals such as *Psychological Review* or *The American Journal of Psychology* published in the 1880s and 1890s illustrates just how much overlap there was between the two disciplines and makes clear that empirical investigations were still being pursued by philosophers despite the standard story. Philosophical topics were under experimental investigation by philosophers, psychologists, and medical practitioners, and included:

- aesthetics (Calkins et al. 1900; Martin 1905)
- belief and knowledge (Sumner 1898)
- corpus analysis of Heraclitus (Patrick 1888)
- epistemological implications of optical illusions (Jastrow 1891; 1892)
- happiness (Brinton 1893)
- logic (Howison 1896; Lloyd 1896)
- memory (Kennedy 1898)

- the longitudinal effect of teaching moral arguments to students (Street 1897)

In fact, so many empirical studies exist from this period, which have largely gone unnoticed in recent times, that it is critical that we return to them if we hope to truly understand what is "traditional" in Western philosophy.

While many philosophers came to embrace methods that have been thought to be independent of empirical investigation, such as conceptual analysis, there was a preoccupation among many empirically minded philosophers that mere speculation would not be sufficient to settle philosophical questions. This is a refrain we see especially in ordinary language philosophy. For example, John Herman Randall (1956) complained that philosophers often do too much talking and not enough looking, a view he had likely inherited from Ludwig Wittgenstein.[1] Let me close this section with a clear example of experimental philosophy in the mid-twentieth century, starting with perhaps the most famous—Arne Næss.

In much of his early philosophical work, Næss promoted an empirical approach to resolving philosophical problems. Næss focused especially on how non-philosophers use ordinary terms of philosophical interest, such as "truth," holding that philosophers can come to a more thorough understanding of a term's usage by engaging ordinary people. In this he recognised that philosophers had been ignoring a valuable research tool.

In this context, Næss points out an array of common phrases that philosophers of the period employed, each of which suggested a body of empirical knowledge that they were not in a position to opine on. These include many of the types of appeals to ordinary or common-sense judgments that contemporary experimental philosophers have frequently targeted, such as "the opinion of the man in the street on the truth-notion is," "to naive people truth means," "[i]f common-sense had been asked to formulate what is meant by the truth of a belief, this is probably what it would have written," and so on (Næss 1938, 14–15). In regard to such appeals, Næss rightly asks, "how do philosophers *know* these things?" Indeed, he notes that "even superficial questioning of non-philosophers makes it hard for anyone to believe that the philosopher has got his 'knowledge' about peasant's and other's use of the word 'true'—or about the views of non-philosophers of truth—by asking any other person than himself" (1938, 40).

In light of this, Næss took an experimental turn, using questionnaires to investigate the accuracy of such claims about the ordinary use of terms like "truth." In these questionnaires Næss asked people a range of open-ended questions, including:

> What is to be understood by the expression "something is true"? Define the expression. (Næss 1938, 24).
>
> What is the c.c. [common characteristics] of that which is wrong? (Næss 1938, 23).
>
> Give me an example of something that is true. (Næss 1938, 23).

[1] As you might expect, a similar sentiment is found among many experimental philosophers today, including that some have taken Wittgenstein's ([1953] 2009, §66) slogan—"don't think, but look!"—as a motivation for their empirical pursuits (e.g., Sytsma and Livengood 2015, 43; Zahorec et al. 2023).

> Do you employ the expression "the truth"? (If answered positively:) On which occasions? (Næss 1938, 26).

Participants' responses to these questions were recorded by an assistant, and the data were then analysed. Even as Næss conducted this work, however, he laments the fact that it was necessary in the first place, noting that much of the work Næss and his lab assistants performed could have been prevented if philosophers had told us of how they came to know what the non-philosopher thinks. Næss writes:

> [T]he fact remains: [Philosophers'] writings contain almost nothing of [how they arrived at the conception of truth among non-philosophers]. Perhaps some of them have asked their wives or assistants for their opinions on the truth-notion, but there is very little to prove that they actually employed such a method. […] Even very superficial questioning of non-philosophers would make it almost impossible for anyone to believe that the philosophers writing about the opinions of ordinary people actually ask others than themselves. (1938, 15).

Since philosophers failed to support their views of the non-philosopher with anything more than mere speculation, Næss believed that a more systematic accounting was needed.

It seems that Næss was conflicted with regard to the ordinary notion. The questionnaire method, though fruitful, did not lead Næss to a uniform view. For instance, he writes that "we have gathered more than 1000 examples from non-philosophers and a great many from philosophic literature, but it is by no means plain how we from this collection should be able to infer any *general* statement resembling definitions" (Næss 1938, 71). Given the wide variety of ordinary notions of truth he had collected using the questionnaire method, none of them stand alone as *the* ordinary notion or commonsense view—a conclusion that prefigures the pluralistic conclusions many contemporary experimental philosophers have arrived at in studying ordinary beliefs (see, e.g., Gonnerman et al 2018, 2021 on knowledge-how, and Ulatowski 2017 on truth).

## 11.3 Experimental Philosophy in the Lvov-Warsaw School: Maria Ossowska

I have shown in §11.2 that experimental methods of x-phi in the last twenty-five years appear to have predecessors in the late nineteenth and early twentieth century. One often hears that analytic philosophy pushed empirically oriented work out of the discipline, starting around the time of Frege and Russell. But, as we've seen, the use of empirical methods occurred in parallel with analytic philosophy, perhaps percolating beneath the surface due to concerns with the "educational crisis" and the value of philosophy outside narrow academic confines. Nonetheless, many philosophers who had become hypnotised by the linguistic turn ignored the experimental work of the late nineteenth and twentieth century, much to the detriment of philosophy, in my view. Only recently has x-phi emerged from the shadow cast upon it by mainstream analytic philosophy, but there are a lot of important, relevant empirical findings that were reported by philosophers in the nineteenth and twentieth centuries, and this work deserves more scrutiny than it has received.

I have already mentioned how the work of philosophers, such as Arne Næss' empirical semantics, may be considered an experimental philosophy because of its employment of survey-based questionnaires on study participants. While I do not want to belabour the point any further here, I want to turn to a different school of thought that arose around the same time that Næss completed his tests on the nature of truth.

In 1895, Kazimierz Twardowski, following his earning a PhD under the tutelage of Franz Brentano and Robert Zimmerman at the University of Vienna, assumed the Chair of Philosophy at the University of Lvov. There, he established a school of thought that lasted until the beginning of the Second World War in 1939.[2]

To show how one of its members, Maria Ossowska, exemplified an experimental approach to philosophical problems, I would like to briefly raise three points about how the environment cultivated by Twardowski contributed to the acceptance of experimental work in philosophy. First, Twardowski believed that philosophical work should be rigorous but clearly articulated. We may take this to mean that any work he or his students were undertaking should be done in a way that analyses a topic or concept deeply, penetrating the idea to its very core, without writing in a way that further obfuscates a point. His view was that clear thinking and clear writing work together. Twardowski writes:

> [S]ome philosophers' unclarity of style is not an inevitable consequence of factors inherent in the subject-matter of their expositions, but has its source in the muddled and vague character of their thinking. The situation would then appear to be as follows: clarity of thought goes hand in hand with clarity of style insofar as whoever thinks clearly would also write clearly, and we would have to conclude that an author who writes unclearly does not know how to think clearly. (1920/1999, 257–258).

Members of Twardowski's Lvov-Warsaw School were the epitome of rigorous and clear thinkers. Leśniewski's profound insights in the foundations of mathematics covered some of the most difficult intellectual terrain; yet, reading even his most technical work is accessible because of its clarity. He deftly argued for a nominalised arithmetic without succumbing to the acceptance of abstracta (*cp.* Simons, 2014).

Second, while what came to be known as a "School," Twardowski did not insist upon his students to take up any specific philosophical view. This is unusual because naming a group a "School" suggests that the de facto creator and leader of it would insist upon proteges and students to take up the views of the School's founder. In fact, by all accounts, Twardowski relished the idea that his students would take up topics and areas that he had not. Here's Izydora Dąmbska on Twardowski's mentorship:

> The philosophers of the Lvov group were not united by any common doctrine, by a uniform world-view. Not the content of philosophy but rather the method of philosophizing and the

---

[2] For an alternative, more inclusive view of who to include in the Lvov-Warsaw School *post*-1939, see Anna Brożek's (2022). According to Brożek, the Lvov-Warsaw School continued to flourish well into, if not beyond, the 1950s.

> common language were the factors which formed the foundation of the spiritual community of those people. This is why the school could produce spiritualists and materialists, nominalists and realists, logicians and psychologists, philosophers of Nature and art theorists. (Dąmbska 1948, 17).

Notably, several members of the Lvov-Warsaw School were accomplished scholars of mathematics, computer science, psychology, and sociology. And, importantly, Twardowski himself, when he was a student of Brentano's at the University of Vienna, was surrounded by colleagues like Carl Stumpf, Anton Marty, and Christian von Ehrenfels who would go on to become prominent in the fields of psychology. Even Brentano wrote on descriptive psychology; his magnum opus, *Psychology from an Empirical Standpoint* (1874/1995), was a remarkable investigation of problems in the philosophy of mind informed by philosophers, scientists, and psychologists. No student of Twardowski's was forced to perpetuate a particular method of approaching philosophical problems.

Finally, and related to the second point, Twardowski impressed upon his students the important role that history, especially the history of philosophy, should play in coming to grips with contemporary philosophical problems. Just as his promotion of methodological pluralism had been inherited from his teacher, Brentano, so too could we say that this profound respect for historical resources came from Brentano too. Brentano himself deeply admired the works of Aristotle. One of Brentano's earliest works was *The Psychology of Aristotle* (1867/1977), for example, and that work laid the foundation of his later work on the thesis of intentionality. In Twardowski's wake, students, like Jan Łukasiewicz, published on Aristotle as well. Łukasiewicz's *On Aristotle's Principle of Contradiction: A Critical Study* (1910) addressed and challenged the long accepted logical principle, and his *Aristotle's Syllogistic from the Standpoint of Modern Formal Logic* (1951) is a classic study of Aristotle's work in logic.

Twardowski's approach highlighted a dedication to clarity and rigour, promotion of a plurality of methods, and respect for history and the history of philosophy. These were three fundamental principles that created a prolific intellectual ecosystem in Lvov and in Warsaw from which many scholars emerged to become leaders in their field. Maria Ossowska set herself apart from colleagues by taking up a fresh new means of addressing the problems of moral philosophy in a more practical context.

Ossowska herself was a student of Jan Łukasiewicz at the University of Warsaw where she completed her PhD thesis on Stoic axiology. Following the completion of her PhD in 1921 she wrote extensively on analytic philosophy in Britain, including 'Ruch filozoficzny w Anglii. Idealiści i realiści. Behawioryzm' [The Philosophical Movements in England: Idealism, Realism, and Behaviourism] (1923/1983), as well as two prominent articles on Bertrand Russell, one a review of his *Analysis of Mind* published in *Przegląd Warszawski* [*Warsaw Review*] (1922/1983) and the other a journal article 'Ontologia Bertranda Russella' [The Ontology of Bertrand Russell] in *Przegląd Filozoficzny* [*Philosophical Review*] (1924/1983).

Henryk Skolimowski, in his work *Polish Analytical Philosophy*, describes Ossowska's work generously and accurately:

> [Ossowska's] field is moral philosophy or, according to her own terminology, the science of morality. She attempts an examination and description of moral phenomena by establishing what people have considered as bad and as good. She presents a dispassionate analysis of various ethical systems, both by exposing their logical backbones, and by illuminating their crucial points, but not in order to evaluate. [...] The main purpose of this sort of study is to reveal what *does take place* rather than what should take place in the field of morality. (1967, 194).

Given this admission by Skolimowski, it seems reasonable to say of Ossowska's work in the science of morals that it likely fits an experimental paradigm. The work does not involve experiments that she herself conducted; instead, Ossowska supported the empirical work that others had undertaken by citing them and by using the results and data as evidence for philosophical conclusions that she reached. It is important here to remind ourselves that not all work in x-phi needs to include empirical studies for it to be included in the literature. On the contrary, allies of experimental philosophy help by endorsing the important argumentative role that empirical work may play in devising philosophic concepts. These allies, like Ossowska, tend to call upon the results of empirical studies to draw philosophical conclusions. While his description of Ossowska's project does not provide much detail of her work, Skolimowski's reporting that Ossowska's work is more descriptive than normative immediately suggests that her work in the science of morals leaned toward the social sciences, especially psychology and sociology.

Some may object by insisting that almost any work in philosophy amounts to a project of x-phi if it cites empirical evidence as a part of a philosophical argument. Even staunch *a priorists* or armchair philosophers—those philosophers who may be counted amongst x-phi's detractors—like Timothy Williamson have called upon empirical data to support their conclusions. Given that this is true, we have to consider a means of identifying authentic allies of x-phi from those who purport to use empirical work in support of their own conclusions without supporting the use of experiments or experimental data in philosophy.

What I contend in this section of the paper is that Ossowska's work is representative of work in experimental moral philosophy, which explores a subset of ethical problems using the methods of x-phi. Mark Alfano et al. (2022) have defined experimental moral philosophy as:

> [the] study of moral intuitions, judgements, and moral behaviour where authors gather data using experimental methods and use these data to substantiate, undermine, reconcile, or revise philosophical theories of moral philosophy.

Following a definition of experimental moral philosophy, Alfano et al. (2022) then describe a case from ancient philosophy where Herodotus recounts a story of Darius, emperor of Persia, who asked two groups of people how much money someone would have to pay them to eat the dead bodies of their fathers and what would it take to persuade group members to burn their fathers after death. The groups responded negatively to both questions; no amount of money would suffice for eating the dead bodies of their fathers and nothing should be spoken of so horrid an act as burning one's father after death. Herodotus uses the results of these "experiments" as evidence against moral relativism from disagreement.

Alfano et al. (2022) cite more contemporary work that exemplifies what experimental moral philosophy is. For example, they refer to Haidt's (2006) work that showed people intuitively believe that eating a deceased pet or engaging in consensual sibling incest was morally acceptable depended on one's socio-economic status and the many studies that have shown moral intuitions are subject to normatively irrelevant situational factors, such as order-effects or framing effects (See Sinnott-Armstrong 2008; Schwitzgebel and Cushman, 2012).

While Ossowska's work in the science of morals is not as provocative as what Herodotus or contemporary experimental moral philosophy report, she calls upon data in support of her philosophic debate. She points out that there are fruitful and interesting empirical results that deserve to be considered a part of moral philosophy.

Let's begin with the fact that Ossowska carefully considers how extraneous individual differences, such as one's physical environment, gender, age, sex, and other "demographic factors," may be a factor in how people reply to a study that asks questions of their views on moral and legal rules. She spends nearly the entirety of Chapter 2 of *Social Determinants of Moral Ideas* (1970, 27–97) working out what "dependent variables" matter for moral evaluation among the potential study participants and why we should consider these variables in our descriptive analysis. When she undertakes this analysis, Ossowska is engaging in a bit of experimental work. Ossowska is considering what features or characteristics of people may matter for endorsing a moral view. Perhaps people who self-identify as a woman may be more inclined to support a *care ethic* approach to applied moral problems. Maybe those people who work in a particular industry or who have so many years of work experience have a jaded or cynical view of management. Whatever the outcomes may be of Ossowska's work, she addresses how these dependent variables may play a part in our moral evaluations. In this way, she is engaging in a bit of experimental hypothesising over how the dependent variable(s) may influence moral views. The chapter reflects that Ossowska was thorough and decisive in how she was going to interpret the collected data.

I am not the first to notice Ossowska's endorsement of, and respect for, the philosophical relevance of empirical studies and the data that come along with it. There have been characterisations of Ossowska's science of morals as a project in the social sciences, and she herself has offered an honest assessment of her empirical work, too. Czesław Porębski (2019), for example, gave a series of lectures on Polish Value Theory, and the focus of his seventh lecture is Ossowska's science of morals. He helpfully divided Ossowska's science of morals into three categories: *the language of morals*, *the psychology of morals*, and *the sociology of morals*. The division of Ossowska's work into these three categories is immensely useful, and I would like to follow Porębski's three-fold division here. I will cite specific examples from Ossowska's work to exemplify each of the three categories.

First, Ossowska's language of morals investigates the linguistic contours of moral evaluation and norms. If the moral use of norms and evaluative terms could be distinguished from other utterances of non-normative character, then it would be possible to distinguish morality from legal norms, customs, or rules of good manner, i.e., etiquette. Lexical features of terms alone do not distinguish between moral and non-moral uses of terms. Think, for example, of the word 'good'. It may be used in a non-normative way, such as "this is a good knife," or in a normative way, such as "Reducing poverty is a good." This suggests, then, that more work needs to be done to understand what we mean by 'moral' and 'non-moral'.

In the first chapter of *Social Determinants of Moral Ideas*, Ossowska identifies five different meanings of the term "moral." She contends that at least two of them are commonly confused. One meaning of the term "moral" refers to forms of moral action that seem to spring from an innate sensation felt in our everyday dealings, and a second meaning of the term "moral" is purely neutral. Moral, when used in a purely neutral sense, may involve one of three different projects: (i) the analysis of the structure of ethical systems, i.e., metaethics, (ii) the identification and evaluation of moral motivation, i.e., what has come to be called "moral psychology," and (iii) what Ossowska called "the sociology of moral life." It is the last that she spends the most time developing in the book. For example, her discussion of morality and knowledge casts a descriptive eye over what societies believe and how these beliefs influence their moral decisions. She cites how some communities have a moral view where the mother of twins and the twins themselves are killed because the collective knowledge of the community believes that the children must have had two fathers and demonstrate the mother's adulterous behaviour, and Ossowska describes the practices of Zuñi Indians that show it to be immoral to emote during ceremonies meant to bring about rain (Ossowska 1970, 74–76). In this discussion, we find Ossowska engaged in looking at the space of language that the term "moral" occupies and how language-users tend to employ it. When she calls upon the space of language, Ossowska uses empirical data drawn from the social scientific literature to make a point about what counts as 'moral'. Ossowska does this in a way that shows the important role that the empirical data play for drawing philosophical conclusions. While she may not design an experiment herself or run it on human participants, Ossowska's use of the data in her own work implies that she is sympathetic with what would be experimental philosophy. She uses this discussion to set up what the rest of that work will outline (See also Ossowska 1980, 3–4).

In that part of her project that considered the language of morals, Ossowska asks about the specific traits of a moral norm or moral evaluation and how those traits have been exemplified by the language that we employ to talk about such behaviour. She considers what the moral sense of words like 'good' or 'bad,' or 'ought' and its contraries, means and how these terms expressed people's attitudes toward moral or immoral behaviour, noting, of course, significant shifts in the meaning of such terms over time. Here is an example of Ossowska speaking of a distinction between different moral norms:

> [A] claim can be considered valid not with reference to certain principles but with reference to some socio-psychological fact, namely, the fact of being widely experienced in the given group. In this case, a norm recommending action *A* would be legal when used by *X* if *X* felt not only an obligation to perform *A*, but also the pressure of claims on the part of his fellow-men, these claims being sufficiently widely shared to be treated as justified and not as matter of individual caprice. (1960, 255).

Elsewhere in that same work she considers how a phrase that we take to be a moral norm expresses a distinct legal norm.

> [A] norm like "Honor thy father and thy mother" would be legal for *X* if *X* considered not only that people are under the obligation of honoring their parents, but also that parents are entitled to require it as something due to them. To be entitled is, in other words, to have the right to do so or, in still other words, to make justified claims. (Ossowska 1960, 252).

The distinctiveness of a legal norm may be derived from the fact that it not only sets out an obligation for a person to follow but also clearly defines an entitlement owed to a beneficiary of such an obligation. In this instance, the parents of a child are entitled to require that a child honour and respect them given this legal Biblical commandment.

The focus of Ossowska's language of morals is the difference between moral and non-moral terms, and part of the difference has to do with how one conceives of the meaning of the terms. Ossowska's use of evidence that is drawn from empirical studies undertaken by social scientists reflects what some scholars broadly sympathetic with experimental philosophy have done in their own work. For example, early work by Stephen Stich (1983, 1990), a long-time enthusiastic supporter of experimental philosophy, made the same kind of moves as Ossowska. Stich's early work favoured *eliminativism*, and, in particular, his (1983) showed how folk psychological terms like "belief" should be replaced with a focus on internal causal mechanisms. For this conclusion, Stich used evidence from scientific psychology.

Second, then, is Ossowska's psychology of morals where she asks what the conceptual sources are of people's moral evaluations. How one conceives of a particular term plays a part in one's understanding of a term. For example, a moral rule may elicit a feeling of obligation in a person because they conceive of that statement as a moral rule. A commentator on Ossowska's work, Ewa Nowicka, describes how the concept of morality plays out in the minds of non-philosophers:

> In her *Sociology of Morals*, [Ossowska] bitterly pointed out that the general concept of morality was all the time only too similar to the notion of the American drugstore. Both notions, morality and the drugstore, cover a multitude of things of diversified origin and character. Ossowska's efforts to outline the concept of morality in the historical–geographical framework or European tradition turned out to be a failure. She was not able to indicate positive criteria that would allow the classification of phenomena as belonging or not belonging to the sphere of morality. (Nowicka 1977, 84).

Nowicka explains how Ossowska likened the use of the concept of morality among non-philosophers as being like that of an "American drugstore" because the concept applies to different things for different people without being applied in a unified way to something in particular.

In her collaborative work, "The Science of Science," with Stanisław Ossowski, she identifies why the psychology of morals ought to be respected by scientists:

> Modern culture enables man in an ever-growing degree to transform his environment according to his own aims instead of adjusting himself to it. In relation to former cultures this is a rather quantitative difference, but in such a powerful degree that it is impossible not to see it a qualitative difference as well. Not limiting itself to ruling over an extra-human

> environment, science endeavours to assume control over human instincts, over social and economic forces. In the last few decades scientific plans have been forming the organisation of human life on a world scale. (Ossowska and Ossowski 1964, 81).

The human instincts spoken of here by the Ossowskis are the concepts that people deploy when they think about the environment and scientific culture which they populate. Scientists believe that they can manipulate how people view the world and what they believe by reporting facts to an audience that are drawn from experiments that they perform. This is what the Ossowskis mean by "assuming control over human instincts." Scientists may exert influence upon a general audience and their conceptual resources. Here again in these two examples, we see how Ossowska believed that people's views may play an evidential role in philosophical debates.

That scientists exert such influence, however, should not take away the fact that people come to conclusions about the motivations for action from a deep-seated familiarity with how we go about performing action. The circumstances surrounding people's evaluations of morally loaded events may have to do with an appreciation of human actions and motivation. In a discussion over human motivations Ossowska cites the interdisciplinary work of Arne Næss:

> In his study called *Objectivity of Norms*, [...] Arne Næss enumerates factors which work for an overestimation of similarities and those which work for their underestimation. Supplementing some of his suggestions with my own, I should like to count among the factors working for overestimation: 1) the tautological character of norms and their vague and general formulation; 2) the egocentric and ethnocentric attitude of the researchers, who impute to others their own reactions and adjust concepts to their own conceptual repertory; 3) the tendency of the persons interviewed to answer questions so as to please the researcher. As factors contributing to the underestimation of similarities I wish to point out: 1) a tendency to see a difference of moral opinions in the difference of customs, although different customs may well be connected with an identity of moral attitudes. Thus different habits of mourning are associated with a common belief that one ought to show grief after the death of one's relatives. 2) An insufficient taking into account that the differences may lie in the strategy for reaching certain goals, rather than in the goals themselves; 3) an insufficient taking into account of the possibility that the difference may be done to different beliefs, while looking for universally accepted moral values we look for differences, in spite of a common factual basis. (Ossowska 1970, 117).

Ossowska identifies the object of moral evaluation as the blameworthiness of agents involved in a morally loaded action. Blame is a kind of resentment and praise is a kind of attitude. She continues: "We would be inclined to regard intentions, decisions, etc., not as something that is evaluated but as something with respect to which the evaluation is formulated" (Ossowska 1970, 117). When someone is acting in a self-sacrificing way without knowing that what he is doing is regarded as good, his moral merit is more praiseworthy. However, when someone is not acting in a self-sacrificing way but a self-aggrandising way, the moral merit of the agent's action is missing. One is acting from egoistic motives, for example.

In this way, Ossowska's views anticipate Knobe's asymmetry, which was mentioned earlier, and suggests a means of analysis: there are pragmatic considerations to take into account when we analyse the intentionality of action. Built directly into the concept of moral action is an individual's pragmatic assessment of the blameworthiness or praiseworthiness of an action. Given that the chairman of the board in the harm scenario acts callously without regard for the environment, people are quicker to judge that the chairman is blameworthy for harming the environment. Equally, given the callousness of the chairman's decision and its egoistic motives, the chairman in the help scenario does not deserve our praise. Ossowska, then, not only undertakes experimental work in the exploration of the language and psychology of morals but also seems to anticipate a potential response to Knobe's (2003) experiment.

Finally, Ossowska's sociology of morals determines factors that play a part in the moral norms of social groups and how they function. In *Moral Norms* (1980), Ossowska identifies at least five different factors that help to determine a social group's morality: the physical environment, biological, demographic, and economic factors, social differentiation, the political organisation of society, and education level, and religiosity. These factors play, according to Ossowska, a role in the origin and development of moral norms by means of moral education and training, and they serve to problematise the legitimacy of universal moral standards that have not been governed or influenced by sociological background.

## 11.4 Challenges

Before I close this chapter of my work justifying the claim that Ossowska's corpus represents an experimental approach to work in moral philosophy, I would like to consider four challenges against my interpretation. The four criticisms may be divided into two broad categories. The first category of criticism contends that Ossowska's work merely appears to be experimental than that it is truly representative of a work in that domain. The second category of criticism attempts to exclude Ossowska from the experimental tradition by showing how the methodologies she employed does not fit with her intellectual upbringing but fits with other disciplines, namely psychology, sociology, and anthropology.

First, a critic may argue that the experiments that Ossowska reviews in her works were predominantly completed by scholars besides herself and that, if the experiments were performed by others, then she herself was not an experimental philosopher. This criticism has merit because on at least one restrictive view of experimental philosophy only those philosophers who create hypotheses and devise experiments count as doing experimental philosophy. Since there are many experiments that Ossowska reports in her work that were not her own, she should not be considered doing experimental philosophy. Ossowska's corpus is vast, and many of the works have not yet been translated into English to be made widely available. However, of the limited part of her corpus at least 50% of what she has written includes empirical hypotheses and evidence from experiments that others performed. One may believe that that alone excludes her from being an experimental philosophy; however, that others have collected the data should not deter us. Think, for example, of experiments done by social scientists who use the data available from the United States' General

Social Survey ("GSS"). The hypotheses are generated by an investigator and then the data are drawn from the GSS. This counts just as much as an experiment as one who collects the data herself. Thus, Ossowska's work cannot be ruled out as being an example of experimental philosophy.

Second, one may claim that I am reading back into the work of an historical figure the existence of a methodology that did not exist at the time she was working. My concern about this criticism is the assumption that there is a strong distinction between analytic philosophy and other approaches that applies to all philosophical work. This distinction ignores the fact that we now have terms available to us that people working at an earlier time did not have. That they did not have those terms available should not preclude them from being identified as a proponent of the method. Philosophical questions and analyses of *that* time were less likely to distinguish so clearly between social scientific approaches and analytic philosophical approaches to questions of philosophical importance. As I mentioned in §2, there were philosophers publishing in psychology journals, and psychologists publishing in philosophy journals. There was at that time no terminological means of distinguishing between the social sciences and humanities. Therefore, if anyone is reading an interpretation back into the work of historical figures it is my critic and not me.

The third and fourth challenges to my view of Ossowska's work involve my misinterpreting the Lvov-Warsaw School's methodology. A critic may raise the possibility that no one in the Lvov-Warsaw School sought to explore philosophical topics through any other method than that which is traditionally associated with the school itself. This criticism would not likely be raised by someone who is familiar with the school. Which method is *the* one that is traditionally associated with the school? There was not just one method traditionally associated with the school; that is precisely what makes the school inherently interesting. Members of the Lvov-Warsaw School from Twardowski onwards were encouraged to explore a variety of means to deal with philosophical problems and those means may include social scientific methods. Hence, there is no reason to believe that this third criticism would have any merit.

Finally, one may argue that Ossowska was not a philosopher but a social scientist, a sociologist or psychologist. If she is not a philosopher, then how she approaches philosophical problems could not align with what experimental philosophy does. There are two responses to this challenge. First, nothing rules out psychologists or sociologists or linguists or whoever from undertaking a project of experimental philosophy. Nothing. People trained in psychology, like, for example, Liane Young (e.g., Young and Dungan 2012) or Jonathan Haidt 2001), may engage in experimental philosophy as much as people trained in philosophy do. A second reply to this criticism is to point out the obvious that Ossowska was, very much in fact, a philosopher by training. She completed her PhD in Philosophy under the supervision of Jan Łukasiewicz on Stoic axiology. It would take an inordinate amount of mental gymnastics to convince us either that Łukasiewicz was not a philosopher or that axiology is not a subdiscipline of philosophy. So, for these two reasons, the fourth criticism challenging my view fails just as much as the first three do.

## 11.5 Conclusion

Ija Lazari-Pawłowska has said of Ossowska that she felt more closely associated with the social sciences than with philosophy, and Ossowska encouraged her students to take an experimental approach to the study of morality. Lazari-Pawłowska writes:

> [Ossowska] tried to teach her students to study moral phenomena as dispassionately and objectively as a botanist observes plants; she taught them not to distort the description of moral phenomena under the impact of their likes and dislikes. [...] According to her, the task of the moral scientist is to describe, classify and explain moral phenomena, but he should refrain from evaluations. (1977, 198).

Ultimately, Lazari-Pawłowska disagrees with Ossowska's approach because she believes it is impossible to approach moral matters in a purely descriptive and objective way that emulates what a botanist or natural scientist does. Lazari-Pawłowska contends that Ossowska's work contains "confessions of her own emotional reactions" amid the heavily sterilised language used to report the results of her experiments. Moreover, Ossowska was likely "guided by her own introspection" when she claimed that there was a "relationship between moral goodness and beauty" (Lazari-Pawłowska 1977, 203).

One may find it surprising that I agree with Lazari-Pawłowska's assessment because I have used this space to show how careful and methodical Ossowska was in the deployment of her empirical 'science of morals'; however, I do not believe that Lazari-Pawłowska's evaluation of Ossowska represents a challenge or criticism of her work. On the contrary, I firmly believe that Ossowska would agree with Lazari-Pawłowska's claim that we cannot approach work, even scientific experiments, in a completely disinterested way. Let's admit that scientists derive hypotheses that they test through experimentation, and these hypotheses are usually derived from hunches that the scientist may have about what the collected data may show. Sometimes the hypothesis is overturned by the data but sometimes it is borne out by the data. What Ossowska would find objectionable, and what I believe any self-respecting scientist would find objectionable, about scientific experimentation is one where the scientist manipulates the data or method of experimentation to draw conclusions that align with the scientist's own view. That is not what Ossowska, or anyone else for that matter, stands for in work on 'the science of morals.'

Contemporary forms of experimental philosophy do not take place in a vacuum. Ossowska's views serve as an insightful foundation from which to undertake more expansive studies to interrogate issues of moral norms and evaluation using empirical methods. What I have argued in this contribution is that Ossowska was a pioneer in experimental moral philosophy whose work has unfortunately been forgotten for one reason or another; may this work serve as a means to reignite interest in the science of morals and expand our discipline's collective interest in Ossowska's work.

**Acknowledgements** I would like to acknowledge the support of the Institute of Philosophy at the University of Łódź for the opportunity to attend the conference on Forgotten Giants of Science, especially Joanna Papiernik, Elżbieta Jung, and Bartosz Żukowski, and members of the audience who contributed to the discussion that followed my presentation, particularly Zofia Hałęza. I am grateful for the feedback I received from audiences at the 2024 Aotearoa New Zealand Association of Philosophers conference and the online conference in early 2025 held through the Lvov-Warsaw School Research Centre. I appreciate the feedback I have received on the paper from colleagues, especially Anna Brożek, Artur Kosecki, Zuzanna Rybaříková, and Cory Wright, as well as Greg Dawes, Lisa Ellis, Michael LeBuffe, Nick Munn, Padraic Tahua O'Leary, Charles Pigden, Vanessa Schoules, Justin Sytsma, Zach Weber, and Dan Weijers. And I am grateful to McMaster University's Department of Philosophy and the Bertrand Russell Research Centre for providing me with the space and resources to complete this project during Northern Autumn 2025.

## References

Adams, Fred, and Annie Steadman. 2004a. Intentional Action in Ordinary Language: Core Concept or Pragmatic Understanding? *Analysis* 64 (2): 173–181.

Adams, Fred, and Annie Steadman. 2004b. Intentional Actions and Moral Considerations: Still Pragmatic. *Analysis* 64 (3): 268–276.

Alfano, Mark, Edouard Machery, Alexandra Plakias, and Don Loeb. 2022. Experimental Moral Philosophy. In *The Stanford Encyclopedia of Philosophy*, Fall 2022 Edition, edited by Edward N. Zalta and Uri Nodelman. https://plato.stanford.edu/archives/fall2022/entries/experimental-moral/

Barnard, Robert, Joseph Ulatowski, and Jonathan Weinbeerg. 2021. The Fourfold Route to Empirical Enlightenment: Experimental Philosophy's Adolescence and the Changing Body of Work. *Filozofia Nauki* 29 (2): 77–113.

Bednarz, Irena. 2022. Bertrand Russell w polskich publikacjach. *Przegląd Filozoficzny—Nowa Seria* 31(4): 53–76.

Bergson, Henri. 1886. De la simulation inconsciente dans l'état d'hypnotisme. *Revue Philosophique De La France Et De L'étranger* 22: 525–531.

Biber, Douglas, Susan Conrad, and Randi Reppen. 1998. *Corpus linguistics: Investigating Language Structure and Use*. Cambridge: Cambridge University Press.

Brentano, Franz. 1867/1977. *The Psychology of Aristotle*. Berkeley, CA: University of California Press.

Brentano, Franz. 1874/1995. *Psychology from an Empirical Standpoint*. London: Routledge.

Brinton, Daniel. 1893. *The Pursuit of Happiness. A Book of Studies and Strowings*. Philadelphia: David McKay.

Brożek, Anna. 2022. The Lvov-Warsaw School after 1950. *Edukacja Filozofczna* 74: 141–160.

Calkins, Mary Whinton, Helen Buttrick, and Mabel M. Young. 1900. Wellesley College Psychological Studies. An Attempted Experiment in Psychological Æsthetics. *Psychological Review* 7(6): 580–591.

Dąmbska, Izydora. 1948. Czterdzieści lat filozofii we Lwowie, 1898–1938. *Przegląd Filozoficzny* 44: 14–25.

Dummett, Michael. 1993. *Origins of Analytical Philosophy*. Cambridge, MA: Harvard University Press.

Gonnerman, Chad, Kaija Mortensen, and Jacob Robbins. 2018. The Ordinary Concept of Knowledge-How. In *Oxford Studies in Experimental Philosophy*, vol. 2, ed. Tania Lombrozo, Joshua Knobe, and Shaun Nichols, 104–115. Oxford: Oxford University Press.

Gonnerman, Chad, Kaija Mortensen, and Jacob Robbins. 2021. KNOWING HOW as a Philosophical Hybrid. *Synthese* 199 (3/4): 11323–11353.

Greene, Joshua D., R. Brian Sommerville, Leigh E. Nystrom, John M. Darley, and Jonathan D. Cohen. 2001. An fMRI Investigation of Emotional Engagement in Moral Judgment. *Science* 293 (5537): 2105–2108.

Haidt, Jonathan. 2001. The Emotional Dog and its Rational Tail: A Social Intuitionist Approach to Moral Judgment. *Psychological Review* 108 (4): 814–834.

Haidt, Jonathan. 2006. *The Happiness Hypothesis: Finding Modern Truth in Ancient Wisdom*. New York: Basic Books.

Hansen, Nat, J. D. Porter, and Kathryn Francis. 2021. A Corpus Study of "Know": On the Verification of Philosophers' Frequency Claims about Language. *Episteme* 18: 242–268.

Howison, George Holmes. 1896. Psychology and Logic. Further Views. *Psychological Review* 3(6): 652–657

Jastrow, Joseph. 1891. Studies from the Laboratory of Experimental Psychology of the University of Wisconsin. *The American Journal of Psychology* 4 (2): 198–223.

Jastrow, Joseph. 1892. Studies from the Laboratory of Experimental Psychology of the University of Wisconsin. *II. the American Journal of Psychology* 4 (3): 381–428.

Kennedy, Francis. 1898. On the Experimental Investigation of Memory. *Psychological Review* 5 (5): 477–499.

Knobe, Joshua. 2003. Intentional Action and Side Effects in Ordinary Language. *Analysis* 63: 190–193.

Knobe, Joshua. 2004. Intention, Intentional Action and Moral Considerations. *Analysis* 64 (2): 181–187.

Knobe, Joshua, and Shaun Nichols, eds. 2008. *Experimental Philosophy.*, vol. 1. Oxford: Oxford University Press.

Knobe, Joshua, and Shaun Nichols, eds. 2017. 'Experimental Philosophy.' In *The Stanford Encyclopedia of Philosophy*, Winter 2017 Edition, edited by Edward N. Zalta. https://plato.stanford.edu/archives/win2017/entries/experimental-philosophy/

Kosecki, Artur. Forthcoming. On Maria Ossowska's Views on the Ontology of Bertrand Russell.

Lazari-Pawłowska, Ija. 1977. Maria Ossowska as a Moral Scientist. *Dialectics and Humanism* 2: 197–208.

Lloyd, Alfred H. 1896. A Psychological Interpretation of Certain Doctrines in Formal Logic. *Psychological Review* 3 (4): 422–426.

Łukasiewicz, Jan. 1910. *O zasadzie sprzeczności u Arystotelesa, Studium krytyczne [On the Principle of Contradiction in Aristotle: A Critical Study]*. Kraków: Akademia Umiejętności.

Łukasiewicz, Jan. 1951. *Aristotle's Syllogistic from the Standpoint of Modern Formal Logic*. Oxford: Clarendon Press.

Martin, Lillien. 1905. Psychology of Æsthetics. I. Experimental Prospecting in the Field of the Comic. *The American Journal of Psychology* 16(1): 35–118.

McCarthy, Michael J., and Anne O'Keeffe, eds. 2010. *The Routledge Handbook of Corpus Linguistics*. London: Routledge.

McEnery, Tony, and Andrew Wilson. 2001. *Corpus Linguistics. An Introduction*. Edinburgh: Edinburgh University Press.

Næss, Arne. 1938. *"Truth" as Conceived by Those Who Are Not Professional Philosophers*. Oslo: Kommisjon Hos Jacob Dybwad.

Nowicka, Ewa. 1977. The Study of Morality and Cultural Anthropology: Moral Problems of Alien Societies in the Works by Maria Ossowska. *Polish Sociological Bulletin* 39–40 (3–4): 81–93.

Ossowska, Maria. 1960. Moral and Legal Norms. *Journal of Philosophy* 57 (7): 251–258.

Ossowska, Maria. 1970. *Social Determinants of Moral Ideas*. Philadelphia, PA: The University of Pennsylvania Press.

Ossowska, Maria. 1980. *Moral Norms*. Amsterdam: North-Holland Publishing Company.

Ossowska, Maria, and Stanisław Ossowski. 1964. The Science of Science. *Minerva* 3 (1): 72–82.

Ossowska, Maria. 1922/1983. Recenzja z: B. Russell, *The Analysis of Mind*. In *O człowieku, moralności i nauce*. Miscellanea, edited by Maria Ofierska and Maria Smoła, 14–20. Warszawa: PWN.

Ossowska, Maria. 1923/1983. Ruch filozoficzny w Anglii. Idealiści i realiści. Behawioryzm. In *O człowieku, moralności i nauce*. Miscellanea, eds. Maria Ofierska and Maria Smoła, 68–73. Warszawa: PWN.

Ossowska, Maria. 1924/1983. Ontologia Bertranda Russella. In *O człowieku, moralności i nauce.* Miscellanea, eds. Maria Ofierska and Maria Smoła, 74–92. Warszawa: PWN.

Ossowska, Maria. 1947. *Podstawy nauki o moralności.* Warszawa: Czytelnik.

Ossowska, Maria. 1949. *Motywy postępowania—Z zagadnień psychologii moralności.* Warszawa: Książka i wiedza.

Ossowska, Maria. 1956. *Moralność mieszczańska.* Łódź: Ossolineum (Wrocław). English edition: Ossowska, Maria. 1986. *Bourgeois Morality.* Translated by G. L. Campbell. London: Routledge and Kegan Paul.

Patrick, George Thomas White. 1888. A Further Study of Heraclitus. *The American Journal of Psychology* 1(4): 557–690.

Porębski, Czesław. 2019. *Lectures on Polish Value Theory.* Leiden: Brill Publishers.

Randall, John Herman Jr. 1956. Talking and Looking. *Proceedings of the Aristotelian Society* 30: 5–24.

Schwitzgebel, Eric, and Fiery Cushman. 2012. Expertise in Moral Reasoning? Order Effects on Moral Judgment in Professional Philosophers and Non-Philosophers. *Mind and Language* 27 (2): 135–153.

Simons, Peter. 2014. Arithmetic in Leśniewski's Ontology. In *The History and Philosophy of Logic in Poland*, ed. Kevin Mulligan, Katarzyna Kijania-Placek, and Tomasz Placek, 227–241. Basingstoke, UK: Palgrave Macmillan.

Sinnott-Armstrong, Walter. 2008. Framing Moral Intuitions. In *Moral Psychology.* Vol. 2: *The Cognitive Science of Morality*, edited by Walter Sinnott-Armstrong, 47–76. Cambridge, MA: MIT Press.

Skolimowski, Henryk. 1967. *Polish Analytical Philosophy: A Survey and a Comparison with British Analytical Philosophy.* New York: The Humanities Press.

Stich, Stephen. 1983. *From Folk Psychology to Cognitive Science: The Case Against Belief.* Cambridge: MIT Press.

Stich, Stephen. 1990. *The Fragmentation of Reason.* Cambridge: MIT Press.

Street, Jacob Richard. 1897. A Study in Moral Education. *The Pedagogical Seminary* 5 (1): 5–40.

Sumner, Francis Bertody. 1898. A Statistical Study of Belief. *The Psychological Review* 5(6): 616–631.

Sytsma, Justin, and Jonathan Livengood. 2015. *The Theory and Practice of Experimental Philosophy.* New York: Broadview Press.

Sytsma, Justin, Roland Bluhm, Pascale Willemsen, and Kevin Reuter. 2019. Causal Attributions and Corpus Analysis. In *Methodological Advances in Experimental Philosophy*, ed. Eugen Fischer and Mark Curtis, 209–238. London: Bloomsbury.

Twardowski, Kazimierz. 1920/1999. On Clear and Unclear Philosophical Style. In *On Actions, Products and Other Topics in Philosophy*, edited by Johannes Brandl and Jan Woleński, 257–260. Amsterdam: Rodopi.

Ulatowski, Joseph. 2017. *Commonsense Pluralism about Truth. An Empirical Defence.* Basingstoke: Palgrave Macmillan.

Wittgenstein, Ludwig. 1953. *2009. Philosophical Investigations. Translated by G. E. M. Anscombe, P. M. S. Hacker, and Joachim Schulte*, 4th ed. Oxford: Wiley-Blackwell.

Woleński, Jan. 2022. Russell i Polacy. *Przegląd Filozoficzny—Nowa Seria*, 31(4): 93–110.

Young, Liane, and James Dungan. 2012. Where in the Brain is Morality? *Everywhere and Maybe Nowhere. Social Neuroscience* 7 (1): 1–10.

Zahorec, Michael, Robert Bishop, Nat Hansen, John Schwenkler, and Justin Sytsma. 2023. Linguistic Corpora and Ordinary Language: On the Dispute between Ryle and Austin about the Use of 'Voluntary', 'Involuntary', 'Voluntarily', and 'Involuntarily'. In: *Experimental Philosophy of Language: Perspectives, Methods and Prospects*, edited by David Bordonaba-Plou, 121–149. Cham: Springer.

**Joseph Ulatowski** is Bertrand Russell Visiting Professor of Philosophy at McMaster University and Senior Lecturer in Philosophy at the University of Waikato. His interest in the Lvov-Warsaw School concerns topics in philosophical logic and metaphysics, especially the theory of truth, and the promotion of women's contribution to the school and the wider philosophic community.

# Chapter 12
# Relationalism about Space and Time: Revisiting an Unfinished Revolution

**Antonio Vassallo**

**Abstract** This essay revisits the Leibnizian/Machian tradition of relationalism in light of recent developments in physics and metaphysics. While historically dismissed as conceptually elegant but technically deficient, relationalism has re-emerged as a viable foundation for fundamental physics through the advent of pure shape dynamics and the metaphysics of self-subsisting structures. These approaches address longstanding philosophical and physical challenges concerning empirical adequacy, the reconstruction of physical magnitudes, and the grounding of dynamical laws. By re-examining the trajectory of relationalist thought and clarifying how recent proposals fulfil its core ambitions, the essay presents a case for the renewed relevance of relationalism in foundational research.

**Keywords** Relationalism · Pure shape dynamics · Self-subsisting structures

## 12.1 Introduction

Few debates in natural philosophy have been as enduring as the one between relational and substantival conceptions of space and time. From Leibniz's early articulation of a relational view to Mach's critique of Newtonian mechanics, relationalism has repeatedly returned as an attempt to understand spatiotemporal structure as emergent from matter and its configurations. Yet for much of the history of physics, this view remained more of a philosophical aspiration than a physical framework.

The perceived failure of relationalism was codified with clarity and historical nuance by John Earman in Earman (1989), where he argued that the relationalist program had consistently fallen short: Lacking precise formulation, failing to reproduce observed phenomena, and offering no serious rival to the explanatory power of frameworks such as Newtonian mechanics or general relativity, which rely on

A. Vassallo (✉)
Faculty of Administration and Social Sciences, Warsaw University of Technology, Warsaw, Poland
e-mail: antonio.vassallo@pw.edu.pl



E. Jung et al. (eds.), *The Forgotten Giants: Overlooked Revolutions in Natural Philosophy*, Studies in History and Philosophy of Science 70,
https://doi.org/10.1007/978-3-032-34008-5_12

spatiotemporal structures that are external to material systems. Relationalism, in Earman's reading, amounted to a metaphysical aspiration in search of a viable physical theory. Furthermore, philosophers such as Shamik Dasgupta have argued that the traditional framing of the debate—cast in terms of modality or representation—should be replaced with a more fine-grained account in terms of metaphysical grounding (Dasgupta 2011). On this view, the central question is whether spatial facts are explained by or grounded in the distribution of matter and its relations, or vice versa. Similarly, Hartry Field has challenged relationalists to account for the full quantitative structure of space and time—the very geometry and metrical content of physics—without positing a substantivalist manifold (Field 1984). These critiques sharpen the stakes: A viable relationalism must not only eliminate surplus structure, but do so in a way that preserves explanatory and descriptive adequacy.

In the last few decades, a series of developments under the umbrella of shape dynamics (Mercati 2018) and pure shape dynamics (Koslowski et al. 2022) have dramatically altered the landscape. Championed by the work of Julian Barbour and Bruno Bertotti (1977, 1982), these frameworks propose a fundamentally relational formulation of dynamics based on reduced configuration spaces, i.e., *shape spaces*, that eliminate any notion of absolute position, orientation, and scale. This reduction is achieved both formally and metaphysically: it corresponds to a commitment to ontological parsimony, where only those features that track physical differences among possible worlds are retained. Moreover, the pure shape dynamics framework has been embedded within a broader metaphysical picture, most notably, the notion of *self-subsisting structures* developed in Vassallo et al. (2022, 2024). These are relational configurations that exist independently of any embedding space or external time, offering a candidate ontology that is both minimalist and dynamically rich. The combined framework of pure shape dynamics and self-subsisting structures introduces new strategies for recovering physical quantities from purely relational information, thus responding in a principled way to concerns about explanatory adequacy and ontological clarity.

This essay examines how these developments represent a genuine step forward for the relationalist program. Section 2 traces the classical roots of relationalism and the conceptual tensions that persisted between its early philosophical ambitions and the structure of physical theories. Section 3 identifies and consolidates the main philosophical challenges that relationalism has faced in modern discourse. Section 4 introduces the frameworks of shape dynamics and pure shape dynamics, with emphasis on the latter's formal and conceptual distinctiveness. Section 5 develops the metaphysical articulation of pure shape dynamics via the theory of self-subsisting structures, including its treatment of time. Section 6 addresses quantum extensions of the framework, highlighting the importance of adapting relationalist principles to quantum theory if relationalism is to offer a comprehensive foundation for modern physics. Section 7 offers a historiographical perspective on the marginalisation of relationalism and its recent revival. Finally, Section 8 reflects on the unfinished aspects of the framework and outlines what remains to be done to fulfil the promise of a truly relational foundation for physics.

The discussion aims to provide a clear and self-contained account of how relationalism, long dismissed as a philosophical dead end, has re-emerged as a viable research program in both physics and metaphysics. Reassessed through this modern lens, it becomes plausible that what we can call an "unfinished revolution" remained so because it awaited the right conceptual tools. Those tools, this essay contends, are now available.

## 12.2 The Classical Tradition

While a reliance on relational motion can be traced back to Cartesian physics, relationalism begins its full-fledged philosophical development in the writings of Gottfried Wilhelm Leibniz, who argued that space and time are not substances or independent entities but abstractions from the order of coexisting and successive material things. In his correspondence with Clarke (Ariew 2000), Leibniz maintained that to speak of space as something in which objects are located is to mistake a system of relations for an entity in its own right. He supported this view with two central principles: The Identity of Indiscernibles and the Principle of Sufficient Reason. According to these principles, if two possible worlds differ only in the overall location or orientation of all matter in space, but are otherwise indiscernible in their internal relational content, then they must be metaphysically identical. Any surplus structure beyond relational facts is metaphysically idle.

However, Leibniz's foundational insight lacked a dynamical counterpart. Indeed, Leibniz provided no mechanism for how motion or inertia could be explained purely in terms of relations, leaving the explanatory advantage to Newton, because he could invoke explanatorily powerful notions such as absolute acceleration relative to a fixed background. This asymmetry between the metaphysical elegance of relationalism and the empirical success of substantivalist mechanics set the tone for the debate for over two centuries.

After Leibniz, other thinkers, including Berkeley and Huygens, questioned aspects of Newtonian dynamics and explored relationalist ideas. However, they too did not propose a complete alternative framework. An especially interesting proposal came from Carl Neumann in the nineteenth century (cf. Barbour 2001, §12.2), who introduced the notion of *Body Alpha*: A hypothetical reference body meant to account for inertial effects without positing absolute space. This idea anticipated the suggestion that inertial structure might be determined by the distribution of mass-energy in the universe, rather than by any fixed background.

It is with Ernst Mach that relationalism returns to prominence with renewed critical force. In his Science of Mechanics (Mach 1883), Mach famously analysed Newton's bucket experiment, in which the concavity of water in a rotating bucket is taken to reveal absolute rotation. Mach questioned whether such acceleration is genuinely absolute, proposing instead that inertial effects arise from the body's relation to the rest of the matter in the universe. This relational view, which came to be known as "Mach's principle" (cf. Barbour and Pfister 1995, for a comprehensive discussion of this principle) draws on the idea that what we call inertial motion is defined relative to the "fixed stars," that is, the bulk matter content of the cosmos. Though Mach's proposal lacked formal equations, it marked an important shift: a move from accepting inertial structure as given to seeking its explanation within a fully relational ontology. Einstein drew inspiration from this vision in developing general relativity, though the final formulation of the theory reintroduced substantival elements through the manifold and metric structure of spacetime.

Thus, from Leibniz to Mach, relationalism remained largely a philosophical critique rather than a concrete alternative to existing physical theories. Besides lacking a worked-out mathematical formalism, relationalism was unable also to provide a systematic account of how observable quantities like distance, duration, and acceleration could be recovered from relational configurations. The conceptual continuity from Neumann's Body Alpha to Mach's fixed stars illustrates the recurring aspiration to anchor inertial structure in the matter distribution of the universe, but the explanatory mechanisms remained elusive.

Henri Poincaré brought this difficulty into sharp focus. In a famous thought experiment (Poincaré 2017, Chapter 7), he imagined a community of astronomers confined to a solar system much like ours, unaware of the broader universe. These scientists, he argued, would eventually find it convenient—perhaps even inevitable—to introduce absolute notions such as the orientation or rotation of space in order to make sense of inertial effects. They might, in fact, hypothesise an invisible reference body like Neumann's Body Alpha, but Poincaré warned that such a postulate would function more like a placeholder than an explanation, i.e., a conceptual artifice introduced to preserve empirical adequacy in the absence of a fully relational account. The true challenge is to avoid artifices like this without sacrificing descriptive power.

## 12.3 Challenges to Relationalism

Before turning to recent relationalist proposals, it is useful to recall the main philosophical objections that have historically cast doubt on relationalism's viability as a physical theory. These objections, articulated most prominently by Earman, Field, and Dasgupta, centre on the concern that relationalism cannot match the empirical precision, explanatory scope, and metaphysical clarity of substantivalist frameworks. At the same time, they help to sharpen the standards that any contemporary version of relationalism, faithful to the legacy of Leibniz and Mach, must aspire to meet.

John Earman's historical and philosophical assessment of relationalism in Earman (1989) offers a particularly influential critique. Drawing on the Leibniz–Clarke correspondence, Earman argues that relationalism, while philosophically attractive, consistently fails to offer the formal resources needed to match the empirical and explanatory achievements of its substantivalist rivals. In Newtonian mechanics, for instance, inertial motion and absolute acceleration are defined relative to a fixed background structure: A spatial frame with absolute positions and a temporal parameter with absolute durations. Earman notes that without comparable resources, relationalism is unable to reconstruct key dynamical features, such as inertial frames or the empirical significance of acceleration.

Take the case of two equal-mass material globes tied together by an inextensible rope and uniformly rotating around their centre of mass in an otherwise empty universe. Because of the centrifugal forces acting on this system, the rope will exhibit a tension. The substantivalist has a clear explanation of this empirically observable effect: In the rotating frame of reference tied to the system, the centrifugal forces arise because of the acceleration, which is due to the change of direction with respect to absolute space. So the facts about the forces acting on the system and the subsequent tension of the rope are accounted for and thoroughly explained by the fact that the system rotates with respect to the absolute inertial reference frame represented by absolute space. This scenario also prompts a plausible prediction: If the system stopped rotating, the rope's tension would go to zero because, at that point, the reference frame tied to the system would overlap with the absolute inertial frame (i.e., the system would be at absolute rest). Note how this entire story can also be given a precise mathematical formulation, which can indeed be found in most elementary physics textbooks.

Can the Leibnizian/Machian relationalist provide an equally powerful account of the system without invoking any external spatial background? The answer is simply *no*, because the vocabulary of the relationalist is too meagre to reconstruct the situation. Indeed, the relationalist can only talk about instantaneous spatial distances between the two globes, but these never change throughout the dynamical evolution. Also, there are no "fixed stars" in this universe to which a change of direction can be referred. Therefore, there is no way the relationalist can account for the rope's tension (or lack thereof). Of course, the relationalist may retort that the whole thought experiment is misguided, given that we have no way to ascertain how the system would really behave in an empty universe. However, this would amount to avoiding the challenge rather than answering to it. Clearly, this issue highlights a deep concern about whether empirical phenomena can be described at all without implicitly reinstating the very structures that relationalism aims to avoid. As long as relationalists remain unable to offer principled, systematic methods for defining inertial motion without reference to an absolute background, Earman concludes, they fall short of the standards set by successful physical theories.

This critique dovetails with Hartry Field's influential argument in Field (1984). Field accepts the attraction of eliminating unobservable structure but warns that doing so demands the recovery of a great deal of sophisticated geometry. A relationalist theory must be able to reconstruct distances, durations, angles, and curvature purely from the relations among material entities. In effect, Field challenges the relationalist to derive the entire metric and affine structure of spacetime from a base of purely relational facts. Without this, he argues, any talk of eliminating space and time risks collapsing into mere slogan. Importantly, Field's critique is forward-looking: He imagines that if such a relationalist reconstruction could be supplied, it would indeed undermine the need for substantival spacetime. But until then, the epistemic and ontological advantages of relationalism remain speculative.

Shamik Dasgupta's challenge is of a more explicitly metaphysical nature. In Dasgupta (2011), he reframes the relational–substantival debate in terms of metaphysical grounding. Rather than focusing on modal or representational equivalence, he asks what grounds the spatiotemporal facts we observe. He maintains that the grounding relation is asymmetric: If relationalism is to succeed, then the facts about geometry and motion must be grounded in purely relational facts about matter. But this is precisely what seems to be missing in standard relationalist theories. As Dasgupta emphasises, traditional relationalism lacks the tools to show how the quantitative content of geometry—facts about length, angle, acceleration, and duration—could be metaphysically grounded in configurations of matter alone. A convincing relationalist ontology must therefore do more than eliminate substantival structures; it must show how the spatiotemporal features of the world arise from, or at least are determined by, relational configurations.

The picture that emerges from these objections is one of deep structural dependence: The successes of physical theories appear to rest on elements that relationalism seeks to reject. The tasks of describing inertial motion, recovering metrical relations, and grounding the geometry of space and time seem to require more than relational configurations have traditionally been able to supply. If relationalism is to move beyond philosophical appeal and provide a robust foundation for physics, it must find a way to meet these interconnected demands on its own terms.

Here comes the turning point in this entire story. As it will be argued in the following sections, some modern attempts at articulating a Leibnizian/Machian framework for physics have now come very close to meeting these demands—much closer, in fact, than any other attempt made in the past.

## 12.4 A New Chance for Relationalism

A significant breakthrough in the development of a modern relationalist framework came with Shape Dynamics (SD). Originally developed by Julian Barbour and collaborators, SD aims to reformulate classical mechanics and general relativity in a way that captures the core relationalist insights of Leibniz and Mach (see Mercati 2018, for a comprehensive textbook on the subject, and Gomes et al., 2011, for the original paper proposing the core insights of SD as a theory of gravity "dual" to general relativity). Its key innovation lies in eliminating the reliance on an external spacetime structure and describing dynamics instead within a reduced configuration space that retains only the physically meaningful relational degrees of freedom.

This reduction begins with the insight that many aspects of a system's configuration—such as its overall position, orientation, and scale—are surplus structure, unobservable and dynamically irrelevant. These are factored out by quotienting the full configuration space by the symmetry group of translations, rotations, and dilations, a procedure that captures the very spirit of the Leibnizian critique of absolute space as a redundant structure. The result is shape space, i.e., a space of configurations defined purely by the internal relations among constituents. Importantly, this operation singles out conformal structure as fundamental, that is, the physical content of a configuration lies in the ratios of distances as encoded in the internal angles, which correspond to empirically observable quantities. Measurements, thus, amount to comparisons between different parts of a configuration, one acting as a unit against which the other is gauged. Importantly, this procedure makes sense only for closed systems, ideally the universe as a whole—a global perspective that echoes Mach's take on physics.

Once absolute structure is eliminated, dynamics can no longer be defined in terms of motion through space or evolution in time. Instead, SD constructs dynamics using a timeless action principle defined on shape space. The central tool here is *best-matching*, which is a variational method to define an intrinsic derivative between configurations by optimally aligning them through the allowed symmetry transformations. This alignment defines a notion of *equilocality*—how parts of one configuration correspond to parts of another—and thereby underwrites a relational definition of change. Geodesic paths on shape space, defined by this intrinsic derivative, represent dynamical evolution without any reference to an external clock or coordinate frame. Most importantly, the dynamics obeys the constraints that the total linear, angular, and dilational momenta of the material content of the universe must be zero—an automatic consequence of declaring *ab initio* that total translations, rotations, and scaling transformations are physically vacuous. This construction, among other things, immediately explains why relationalists should regard scenarios like the rotating globes thought experiment as unphysical: Their very setup involves a violation of the global relational constraints (e.g., in a universe inhabited by just two material globes, it makes no sense to define rotational motion with respect to their centre of mass).

To describe geodesic motions over shape space, however, SD still requires a monotonically increasing parameter along each dynamical trajectory. While this parameter is formally arbitrary, it is not dynamically inert: It *must* be specified to compute evolution, yet lies outside the relational data. This residual dependence on an external "time-like" parameter reveals a limitation in SD's formulation. It motivates the transition to Pure Shape Dynamics (PSD), a refinement of the framework that removes the need for parametrisation altogether (cf. Koslowski et al. 2022, for a technical introduction to the framework).

PSD's key innovation is to define a system's evolution through a set of geometric degrees of freedom internal to each curve in shape space. These degrees of freedom function analogously to state variables in thermodynamics: They determine how the system evolves, but are themselves part of the system rather than external to it. In thermodynamics, state variables are large-scale properties that provide a macroscopic description of a system—global characteristics, in this analogy, that summarise the overall state of the relational configuration. As such, the equations of motion in PSD resemble equations of state, describing how a system changes as a function of its internal geometric characteristics. This formulation renders PSD remarkably versatile. The framework is general enough to in principle describe a wide range of physical systems—from N-body classical models to general relativistic systems and even quantum systems—by enriching the intrinsic structure of its dynamical curves as needed.

A central concept in PSD is that of *complexity*: A measure of the degree of clustering or structure in a configuration (see Barbour et al. 2014, for the first articulation of this notion). In an N-particle system, for example, configurations where all particles are equidistant exhibit low complexity, while configurations with highly localised clusters have high complexity. Complexity provides a natural "time stamp" on configurations, allowing one to define a meaningful, intrinsic ordering of configurations without appealing to an external parameter. As complexity increases (or decreases), one can track the system's evolution entirely from within.

Moreover, PSD explains how effective spatial and temporal measurements—rods and clocks—can emerge locally from the global relational dynamics. Stable subsystems, such as Kepler pairs in an N-body system, act as dynamically generated standards. A Kepler pair consists of two particles that become gravitationally bound and trace regular, typically elliptical orbits around their shared centre of mass. Because these motions are dynamically stable and repetitive, they provide a reliable reference for assigning spatial distances and time intervals. Crucially, all Kepler pairs that form throughout the evolution of the system will agree on these standards, since they emerge from the same global dynamical mechanism. In this way, robust local spatiotemporal notions arise consistently and coherently across configurations, all within a theory that is globally spacetimeless—reinforcing the Machian view that the universe as a whole determines local physical content. This feature also speaks to a key epistemological virtue of the framework: It shows how observer-accessible data—such as durations and distances—can be recovered from within a theory that posits no fundamental temporal or spatial background. Instead of assuming rods and clocks as primitive, the framework explains how these emerge dynamically and relationally from the system itself. By accounting for how its own users, as physical subsystems, can access and interpret observable quantities, the theory thus exhibits a remarkable epistemic transparency.

Yet the formal framework of PSD, while powerful, remains incomplete without a metaphysical account of what its structures represent. The theory shows how dynamics can unfold in a purely relational configuration space, but it does not by itself answer the deeper question of what grounds the physical reality of these configurations. To address this, the formalism must be paired with a metaphysical interpretation. That interpretation is provided by the theory of self-subsisting structures.

## 12.5 Self-Subsisting Structures and the Grounding of Relational Dynamics

The theory of self-subsisting structures offers a metaphysical framework that complements PSD, allowing it to meet the philosophical demands outlined in Section 3. Developed primarily in Vassallo et al. (2022, 2024), this framework provides a rigorous ontology for a world fundamentally described by relational configurations, without reference to external spacetime or absolute physical magnitudes.

At the core of this approach lies a strong ontological thesis: The fundamental constituents of reality are not particles, fields, or spacetime points, but entire configurations of matter in relational terms. These configurations are self-subsisting in the sense that they do not exist "in" anything else—they are not embedded in space or evolving in time. Rather, they are complete, self-contained structures defined entirely by the internal relations that underlie a conformal structure—that is, relations specifying the intrinsic ratios of spatial configurations, such as relative angles, which remain invariant under changes of absolute scale. This conception draws from the metaphysical stance known as *ontic structural realism*, which holds that structure—not individual objects—is ontologically primary (see Ladyman 1998, for one of the first modern proposals in this direction). The self-subsisting structures framework provides a concrete physical realisation of this thesis, portraying these relational configurations as the true ontological units of the universe. This shift from individual objects to holistic configurations is tied directly to the formal structure of PSD, where each point in shape space represents a configuration stripped of absolute structure. The self-subsisting structures framework interprets these points as ontologically robust entities. A self-subsisting structure is, in this sense, a possible state of the universe: a complete and autonomous node in the global mosaic of reality. The dynamics of such structures unfolds through intrinsic ordering relations among configurations. The ordering can be grounded in features such as increasing complexity, which functions as a kind of "internal clock." Crucially, this relational evolution arises from lawful relations between structures, encoded in the intrinsic geometry of shape space.

One of the principal advantages of this framework is its explanatory strength. It allows relationalists to answer a version of the grounding challenge posed by Dasgupta: If spatial and temporal facts are derivative, what grounds them? In this framework, such facts supervene on the pattern of configurations and their interrelations. This move builds on the Humean strategy, initially articulated by David Lewis, according to which all facts about laws, causation, and modality supervene on the total distribution of local qualities across spacetime—the so-called *Humean mosaic* (see Lewis 1986, viii–xi, for a brief characterisation of this stance). On this view, the laws of nature do not determine the physical happenings, but rather emerge from the best systematisation of the mosaic: the set of regularities that balance simplicity and strength in describing the world. More recently, Super-Humeanism has extended this picture by eliminating even local qualities as fundamental, treating only the distribution of matter as ontologically basic, and deriving everything else, including geometry and dynamical structure, from this sparse base (Esfeld et al. 2018).

The self-subsisting structures framework pushes this line of thought further. Instead of presupposing a spacetime manifold with localised elements, it takes entire relational configurations as primitive. The mosaic here is not a pattern of local instantiations in space and time, but a sequence of self-subsisting, unembedded structures, each defined solely by internal relational features. Geometry and temporality are recovered as systematisations over this sequence. For instance, distances and durations are grounded in the recurrence and regularity of certain structural motifs—such as Kepler pairs and clustering patterns—while temporal ordering arises from intrinsic complexity gradients between configurations (cf. Vassallo and Naranjo 2025, for the full articulation of such a construction). Thus, spatial and temporal facts are not strictly speaking eliminated, but grounded: They are higher-level constructs that emerge from the intrinsic relations among the self-subsisting structures themselves.

This perspective also offers a response to Field's challenge: How to recover the quantitative structure of physics from a relational ontology. In the self-subsisting structures framework, material patterns license the reconstruction of inertial frames, durations, and metrical structures as systematisations over the relational data. Moreover, dynamical quantities such as accelerations, angular momentum, and other motion-related properties can be similarly reconstructed for subsystems of the universe. This is achieved without compromising the core relationalist commitment, since the global constraints embedded in the theory ensure that the total linear, angular, and dilational momenta of the universe remain exactly zero by construction. In this way, self-subsisting structures enable a consistent and explanatory recovery of dynamical content from a purely relational base.

The framework also resolves a deep tension in the relationalist project, namely, the need for descriptive adequacy without ontological inflation. Because self-subsisting structures are holistic and dynamically interrelated, they can account for physical phenomena without invoking unobservable background entities. Their evolution is tracked by reference to their intrinsic features, rather than an external time. The resulting picture is one in which the laws of physics are patterns across a space of possible relational configurations, rather than prescriptions for how things change in time.

In this sense, this modern relational framework substantially delivers on the original Leibnizian and Machian promise. It provides a formulation of dynamics free of external spatiotemporal structure, but with the metaphysical resources to ground the geometry and temporality that such standard dynamics presupposes. Relational facts are not explained by embedding them into space and time; instead, space and time are emergent aspects of the structure and order among relational facts. The payoff is remarkable: A coherent and physically informed relational ontology that can underwrite the descriptive and explanatory practices of physics without lapsing into substantivalist assumptions.

## 12.6 Quantum Extensions

Any claim to offer a fundamental theory of physics must confront the quantum domain. The empirical success of quantum mechanics, particularly in exhibiting non-local correlations that defy classical expectations (as experimentally proven in the seminal work of Aspect et al. 1981, 1982), compels any metaphysical framework to either accommodate quantum phenomena or risk irrelevance. In this light, the development of a quantum counterpart to the relational framework sketched so far is a clear necessity. Here, however, the conceptual terrain grows more difficult. Quantum mechanics is already philosophically fraught, and its standard formulation is deeply wedded to abstract configuration spaces and wave functions, entities that appear to resist a straightforward relationalist interpretation (or any interpretation, for that matter). To put it in Tim Maudlin's words (Maudlin 2019, 77, emphasis in the original): "What the quantum recipe does not resolve, what it does not even purport to address, is *what the physical world is like such that the quantum recipe works so well.*"

Against this backdrop, it is plausible to assume that a viable quantum extension of PSD should draw from any theoretical approach to quantum physics that clearly specifies the physical nature of quantum systems—i.e., what they are "made of" from an ontological viewpoint—and the underlying dynamical mechanisms that shape them. In this sense, a promising route lies in the de Broglie–Bohm theory (dBB; see, e.g., Bohm 1952a, 1952b, for one of the first fully-fledged attempts at articulating this theory). In its standard formulation, the theory posits N material particles having definite positions at all times; in this picture, the particles' trajectories are determined by the wave function, whose evolution is described by the Schrödinger equation. The way the wave function determines such trajectories is encoded in a mathematical expression suggestively called "guidance" equation. Importantly, this guidance is holistic: The velocity of each particle depends on the wave function, which, in turn, depends on the position of all the particles at a given time. Thus, dBB naturally assumes a global viewpoint, describing the dynamical evolution of a universal configuration of N particles. In doing so, it offers a dynamical account of quantum non-locality that is rooted in the structure of the whole system: All particles instantaneously "feel" each other at any given moment, irrespective of how distant in space they are.

This global perspective makes dBB an ideal candidate for the quantum extension of PSD. Indeed, Vassallo (2015) argues that dBB, although formulated over a spatiotemporal background, is amenable to the quotienting out procedure that leads to a totally relational description of the universe. Furthermore, Farokhi et al. (2024) show that, when the dBB dynamics is formulated within the PSD framework, a striking result emerges: The guiding equation breaks down at the universal level. This means that the universe as a whole is not "quantum" in a dBB sense. Instead, quantum dynamics arises only at the level of subsystems, which elegantly ties with the notion of the *effective* wave function, a construct that encodes how a given subsystem evolves relative to the configuration of the rest of the universe (cf.

Dürr et al. 1992, for a technical introduction to this notion in standard dBB theory). In this way, quantum behaviour becomes an emergent, context-dependent feature of relational subsystems, rather than a fundamental aspect of universal dynamics. In this sense, the PSD approach to dBB theory not only recovers the standard account of how quantum subsystems interact (e.g., during measurements in the lab), but it does so by implementing a core tenet of relationalism, namely, that the dynamical description of the full system requires no external reference structure—a universal guiding wave, in this case.

This shift also bears metaphysical significance. When combined with the framework of self-subsisting structures, the PSD approach to dBB theory supports a view in which quantum dynamics is not grounded in a primitive, universal wave function because it instead supervenes on the sequence of self-subsisting configurations and their internal relations. In this setting, the wave function represents a codification of patterns in the sequence of relational structures. It emerges from the best systematisation of the evolution of relational facts, echoing the Super-Humean strategy discussed in the previous section. Interestingly enough, this perspective seems to go in the direction of recent conjectures in high-energy physics—such as the ER=EPR proposal (Van Raamsdonk 2010)—that suggest a deep connection between quantum entanglement and spacetime geometry. On this view, entanglement may be the fabric from which spatial structure arises. In the self-subsisting structures framework, this relation is made precise: The internal relational structure of configurations grounds both spatial geometry and quantum correlations, that is, space and entanglement are different aspects of the same underlying relational reality (a full philosophical defence of this view can be found in Vassallo et al. 2024, §5).

Moreover, although dBB theory serves as a natural vehicle for developing quantum relationalism, it is by no means the only one. As pointed out earlier, the essential feature required is not determinism or particle trajectories, but rather a clear ontology—some specification of what exists and how it behaves. Any interpretation of quantum theory that provides such a specification—be it an Everettian model, a GRW-type collapse theory, or related approaches—can, in principle, be reformulated in relational terms: What matters is the commitment to relational degrees of freedom as fundamental.[1] In this light, the quantum extension of PSD and self-subsisting structures opens a pathway toward addressing the problem of quantum gravity. By showing that both quantum and classical dynamics can emerge from the same underlying relational base, this framework suggests a route toward unifying these domains. While a full theory of quantum gravity remains elusive, the approach discussed here

[1] An alternative extension of the Barbour–Bertotti framework to the quantum domain has been also developed by Gryb and Thébault (see, e.g., Gryb and Thébault, 2012, 2016). Their program, known as *relational quantization*, treats time-reparametrization symmetry as physically meaningful and recovers a Schrödinger-type evolution for the whole universe. In contrast to the PSD/self-subsisting structures framework, which embraces timelessness at the fundamental level and derives temporal ordering from internal structural features such as complexity, the Gryb–Thébault approach preserves a fundamental quantum dynamics driven by an emergent temporal parameter. While both frameworks aim to implement relationalism, they clearly differ significantly in their treatment of time and the metaphysical commitments this entails.

points in a promising direction: Rather than quantising spacetime, we may understand spacetime and quantum correlations alike as emerging from a deeper, more primitive relational order.

These developments mark a significant departure from the long-standing view that relationalism lacks the resources to confront modern physics. Rather than trailing behind the dominant paradigms, relational frameworks such as PSD—especially when augmented by the metaphysics of self-subsisting structures—have begun to anticipate and recast some of the most pressing questions at the foundations of physics. Yet, despite these conceptual and technical advances, relationalism has remained at the margins of mainstream discourse. Understanding how and why this happened requires taking a step back.

## 12.7 The Relationalist Program in Retrospective

Despite its deep philosophical pedigree, relationalism has long occupied a marginal position in the history of modern physics. From Newton's decisive formulation of absolute space and time, through the geometrisation of gravitation in general relativity, the dominant frameworks of physical theory have consistently relied on an external spatiotemporal structure—be it a fixed background or a dynamical field-like construct. Even where relational ideas were entertained by figures like Leibniz, Mach, and Poincaré, they typically appeared as speculative counterpoints to more mathematically tractable, empirically fruitful substantivalist models.

Several factors contributed to this marginalisation. One is conceptual inertia: once embedded in the mathematical formalism, external spatiotemporal structures acquired a default legitimacy, and the burden of proof shifted to the relationalist to explain how their absence would not cripple explanation or predictive power. Another factor is technical: Until quite recently, relationalism had no framework on par with Newtonian mechanics or general relativity in terms of quantitative precision and empirical applicability. Even attempts like the original Barbour–Bertotti model (Barbour et al. 1982), despite their elegance, remained largely confined to idealised systems and lacked the metaphysical articulation needed to be taken seriously beyond niche circles. Moreover, the philosophical discourse surrounding relationalism often suffered from a disconnect between metaphysical ambition and physical feasibility. Many relationalist arguments remained at the level of principle or intuition, failing to produce a viable, working alternative to mainstream physical theories. Critics like Earman and Field were right to insist that mere conceptual coherence is not enough: A relationalist theory must do real explanatory work if it is to be more than a metaphysical gloss.

Yet in light of the developments surveyed in this essay, this verdict now seems increasingly premature. The combined framework of PSD and self-subsisting structures offers a version of relationalism that is both formally robust and philosophically mature. It provides a concrete formalism in which external spatiotemporal structures are dispensed with, not by fiat, but through principled constraints. Indeed, the framework shows how such structures can emerge from an underlying network of relations. Furthermore, it delivers a technically informed approach to quantum physics that is in line with both empirical data and cutting-edge ideas in foundational physics. In this light, the long marginalisation of relationalism may be understood as the natural consequence of its premature articulation. Lacking the formal and conceptual tools to express itself rigorously, relationalism was bound to remain a philosophical undercurrent rather than a scientific contender. What has changed is not the aspiration but the machinery. Today, that machinery, shaped by the joint efforts of physicists and philosophers, has begun to catch up. Relationalism may thus seriously aspire to represent the next stage in our understanding of physical reality.

## 12.8 Conclusion

This essay has argued that the Leibnizian/Machian vision of a relationalist foundation for physics has in recent years gained new traction through the synthesis of PSD and the metaphysics of self-subsisting structures. This framework overcomes many of the classical objections to relationalism and offers a fertile perspective on time, geometry, and the emergence of quantum behaviour. Yet, despite these advances, the picture remains incomplete. Several key issues, both physical and metaphysical, must be addressed for the relationalist program to reach full maturity.

From a physical standpoint, the most pressing loose end concerns the precise conditions under which quantum behaviour emerges within the PSD-based dBB model. While the absence of a universal guiding equation is conceptually appealing, the theory currently lacks a detailed account of what triggers the emergence of the guidance equation at the level of subsystems. Without such a specification, the framework cannot fully explain what renders the world we observe "quantum." This gap points to the need for a richer characterisation of the pre-quantum regime: A domain where PSD functions as an autonomous theory, not merely as a "rewriting" of quantum or classical mechanics in relational terms. In this primordial regime, where spatiotemporal and quantum behaviour is entirely absent, PSD must define its own observables, kinematical principles, and predictive resources. A related challenge concerns the generalisation of the framework to encompass the degrees of freedom found in realistic field-theoretic contexts. Although some preliminary efforts have been made to provide a treatment of dynamical geometry (Farokhi et al. 2025), the current PSD formalism does not yet fully account for internal symmetries, gauge structure, or matter fields. A mature relational theory must articulate how such features arise from its intrinsic variables, lest it fall short of describing the full content of modern physics.

These physical gaps lead naturally into more deeply metaphysical ones. Chief among them is the question of what metaphysically binds self-subsisting structures together. As discussed, these structures are conceived as ontologically autonomous relational configurations, each encoding a conformal geometry via internal angles.

But the metaphysical basis for their connectedness—what makes them integrated wholes of coexisting relata—remains elusive. Simply postulating relations like spatial proximity or entanglement between them would amount to little more than placing placeholders where a deeper explanation should be. One promising avenue involves reading off a more fundamental dependence relation from fundamental physics, like the dynamical structure of general relativity. Early results suggest that such a relation may capture the co-dependence between matter and geometry that characterises general relativistic physics, offering a potential blueprint for a pre-spatiotemporal "glue" (Vassallo 2020; Vassallo and Hoefer 2020).

Another underdeveloped aspect concerns the articulation of the primitive ideology of the framework. In other words, what are the basic representational resources PSD and self-subsisting structures rely on? Are internal angles and ratios sufficient to describe the full range of relational facts? What logical operations or modal notions are required to speak meaningfully about lawful transitions from one self-subsisting structure to another? These questions press for a more formally rigorous and philosophically perspicuous articulation of the framework's representational foundations. Likewise, further work is needed to clarify whether the framework is best understood as providing metaphysical grounding, scientific reduction, or both. Earman and Field, as we have seen, challenge the relationalist to account for the quantitative structure of physics in a way that preserves its descriptive and explanatory power. The relationalist approach discussed in this essay seems concretely capable of meeting this challenge, but a clearer characterisation of its reconstruction methodology and its metaphysical significance would enhance its dialectical strength. Lastly, the question of temporal ordering and agency remains somewhat obscured in the current formulation. While scalar parameters such as complexity provide a basis for ordering configurations, the framework at the moment lacks a robust account of how we get to perceive time. Given that one of the ultimate goals to secure philosophical dignity for the framework is for it to recover the experience of temporal passage, further metaphysical elaboration will certainly be needed.

These are not fatal flaws but open problems—and the existence of such problems should be expected in any research program that aspires to rethink the foundations of physics. That the PSD and self-subsisting structures framework brings us this far is already a remarkable achievement. However, the recurring caveat of this essay is that, if relationalism is to become a fully mature alternative to substantivalist metaphysics, it must reject the scaffolding of absolute space and time and also supply a coherent and complete account of the structure that remains. As we have seen, such an account really seems within reach: The task ahead is to fully grasp it and to see how far it can take us.

**Acknowledgements** I would like to thank the Editors of this volume for their kind invitation to contribute. I also gratefully acknowledge financial support from the Polish National Science Centre, grant No. 2023/49/B/HS1/01091.

## References

Ariew, R. (ed.) 2000. *G. W. Leibniz and Samuel Clarke - Correspondence*. Hackett Publishing Company, Inc.

Aspect, Alain, Philippe Grangier, and Gérard Roger. 1981. Experimental tests of realistic local theories via Bell's theorem. *Physical Review Letters* 47 (7): 460–463.

Aspect, Alain, Philippe Grangier, and Gérard Roger. 1982. Experimental realization of Einstein-Podolsky-Rosen-Bohm *Gedankenexperiment*: A new violation of Bell's inequalities. *Physical Review Letters* 49 (2): 91–94.

Barbour, Julian. 2001. *The discovery of dynamics*. Oxford University Press.

Barbour, Julian B., and Bruno Bertotti. 1977. Gravity and inertia in a Machian framework. *Il Nuovo Cimento B* 38 (1): 1–27.

Barbour, Julian B., and Bruno Bertotti. 1982. Mach's principle and the structure of dynamical theories. *Proceedings of the Royal Society A* 382:295–306.

Barbour, Julian, Tim Koslowski, and Flavio Mercati. 2014. Identification of a gravitational arrow of time. *Physical Review Letters* 113:181101.

Barbour, Julian B., and Herbert Pfister (eds.) 1995. *Mach's principle: from Newton's bucket to quantum gravity*. Birkhäuser.

Bohm, David. 1952a. A suggested interpretation of the quantum theory in terms of "hidden" variables I. *Physical Review* 85:166–179.

Bohm, David. 1952b. A suggested interpretation of the quantum theory in terms of "hidden" variables II. *Physical Review* 85:180–193.

Dasgupta, Shamik. 2011. The bare necessities. *Philosophical Perspectives* 25:115–160.

Dürr, Detlef, Sheldon Goldstein, and Nino Zanghì. 1992. Quantum Equilibrium and the Origin of Absolute Uncertainty. *Journal of Statistical Physics* 67:843–907.

Earman, John. 1989. *World Enough and Space-Time. Absolute versus Relational Theories of Space and Time*. Cambridge, MA: MIT Press.

Esfeld, Michael, Dirk-André Deckert, Dustin Lazarovici, Andrea Oldofredi, and Antonio Vassallo. 2018. *A Minimalist Ontology of the Natural World*. New York: Routledge.

Farokhi, Pooya, Tim Koslowski, and Pedro Naranjo. 2025. Pure Shape Dynamics: Relational General Relativity. *arXiv:2503.00996v1*. (2 Mar.): https://doi.org/10.48550/arXiv.2503.00996.

Farokhi, Pooya, Tim Koslowski, Pedro Naranjo, and Antonio Vassallo. 2024. A de Broglie–Bohm Model of Pure Shape Dynamics: *N*-Body system. *Foundations of Physics* 54: 40.

Field, Hartry. 1984. Can We Dispense with Space-Time? *PSA: Proceedings of the Biennial Meeting of the Philosophy of Science Association, vol.* 2:33–90.

Gomes, Henrique, Sean Gryb, and Tim Koslowski. 2011. Einstein Gravity as a 3D Conformally Invariant Theory. *Classical and Quantum Gravity* 28 (4): 045005.

Gryb, Sean, and Karim Thébault. 2012. The Role of Time in Relational Quantum Theories. *Foundations of Physics* 42 (9): 1210–1238.

Gryb, Sean, and Karim Thébault. 2016. Schrödinger Evolution for the Universe: Reparametrization. *Classical and Quantum Gravity* 33 (6): 065004.

Koslowski, Tim, Pedro Naranjo, and Antonio Vassallo. 2022. Pure Shape Dynamics: General Framework. *Classical and Quantum Gravity* 39 (21): 215017.

Ladyman, James. 1998. What is Structural Realism? *Studies in History and Philosophy of Science* 29 (3): 409–424.

Lewis, David. 1986. *Philosophical Papers*. Vol. 2. New York: Oxford University Press.

Mach, Ernst. 1883. *Die Mechanik in ihrer Entwickelung, historisch-kritisch dargestellt*. Leipzig: Brockhaus. English edition: Mach, Ernst. 1893. The Science of Mechanics. Translated by T. J. McCormack. Chicago: Open Court Publishing.

Maudlin, Tim. 2019. *Philosophy of Physics: Quantum Theory*. Princeton: Princeton University Press.

Mercati, Flavio. 2018. *Shape Dynamics*. Oxford: Oxford University Press.

Poincaré, Henri. 2017. *Science and Hypothesis. Eds. Mélanie Frappier and David J. Stump. Translated by Mélanie Frappier, Andrea Smith, and David J. Stump.* London: Bloomsbury. First edition 1902.

Van Raamsdonk, Mark. 2010. Building Up Spacetime with Quantum Entanglement. *General Relativity and Gravitation* 42:2323–2329.

Vassallo, Antonio. 2015. Can Bohmian Mechanics be Made Background Independent? *Studies in History and Philosophy of Science Part B: Studies in History and Philosophy of Modern Physics* 52:242–250.

Vassallo, Antonio. 2020. Dependence Relations in General Relativity. European Journal for Philosophy of Science 10 (2). https://doi.org/10.1007/s13194-019-0265-5

Vassallo, Antonio, and Carl Hoefer. 2020. The Metaphysics of Machian Frame–Dragging. In *Thinking About Space and Time*, Einstein Studies, edited by Claus Beisbart, Tilman Sauer, and Christian Wüthrich, vol. 15, 227–246. Cham: Birkhäuser.

Vassallo, Antonio, and Pedro Naranjo. 2025. Pure Shape Dynamics, Self-Subsisting Structures, and the Nature of Time. *Journal for General Philosophy of Science* 56:253–273.

Vassallo, Antonio, Pedro Naranjo, and Tim Koslowski. 2022. A Proposal for a Metaphysics of Self-Subsisting Structures. *I. Classical Physics. Synthese* 200:374.

Vassallo, Antonio, Pedro Naranjo, and Tim Koslowski. 2024. A Proposal for a Metaphysics of Self-Subsisting Structures. *II. Quantum Physics. Foundations of Physics* 54:65.

**Antonio Vassallo** is an Associate Professor of Philosophy at the Warsaw University of Technology, where he coordinates the research group in the philosophy of physics. He previously held academic positions at the Universities of Barcelona, Lausanne, and Warsaw. His research lies at the intersection of metaphysics and fundamental physics, with a focus on the nature of spacetime in general relativity and quantum gravity, as well as on interpretational issues in quantum theory. More recently, his work has focused on modern relational approaches to space, time, and dynamics. He is the editor of The Foundations of Spacetime Physics: Philosophical Perspectives (Routledge, 2022) and has published widely on the philosophy of spacetime and the foundations of physics.

# Index




E. Jung et al. (eds.), *The Forgotten Giants: Overlooked Revolutions in Natural Philosophy*, Studies in History and Philosophy of Science 70,
https://doi.org/10.1007/978-3-032-34008-5

FSC
www.fsc.org
MIX
Papier | Fördert gute Waldnutzung
FSC® C083411

Kolibri 360 GmbH
Ferdinand-Jühlke-Straße 7
99095 Erfurt, Deutschland
produktsicherheit@kolibri360.de